The Tornado Engine +

By: Jim Quinn BSc FIMechE, Design Engineer

MRCA-Tornado

BS143 Concept

RB199-34R

Customer Support

Control Systems

Headway

Remap

Humanism

UK Constitution

European Union

Nuclear Weapons

Software

Evolution

Printed in the United States of America.

Library of Congress Control Number: 2020922232

ISBN	Paperback	978-1-64803-887-7
	eBook	978-1-64803-889-1

Westwood Books Publishing LLC
11416 SW Aventino Drive
Port Saint Lucie, FL 34987

www.westwoodbookspublishing.com

THE TORNADO ENGINE +

A memoir by Jim Quinn BSc FIMechE, Design Engineer

PREFACE

This is the inside story of a man who worked in Rolls Royce Bristol aero engines, for he outlines the work he and others carried out over 21 years during the early concept, development and in service stages of the Panavia MRCA-Tornado engines. He went on to Project Manage Rolls Royce and Europe's first production single crystal turbine blades, and then worked on Control Systems for several years, describing these works as the story proceeds.

A whistle blower, he enjoyed Gilbert & Sullivan, Hang Gliding, and Motorsport, and he eventually left RR, to work with several charities, particularly Headway and Remap, where he describes his experience as an inventive carer of disabled people of all sorts. Many years were spent maintaining and renovating Church Turret Clocks, and some of this is reported here.

As a Humanist, he has experience of and interest in many things, and the book thus contains a variety of challenging thoughts, including the nature of biological evolution (Extra Terrestrials too), a proposal for a suitable UK Constitution because nobody knows the hidden one, many other Proposals for Government, and thoughts about the European Union and nuclear weapons.

CONTENTS

INTRODUCTION

My CV — Jim Quinn BSc FIMechE, Design Engineer

Key Skills

An enthusiastic Professional Engineer used to project management and engineering prediction – a Specialist Manager forming good relationships with all, with a background in Project Concepts, Budgetary Control, Customer Support, Organisation, Software, Control Systems and Auditing.

Computer literate with much experience of internet and email, Windows XP, LibreOffice, PowerPoint and presentations, Z80 Assembler, Scheme P, Fortran, Visual Basic, C++, 2D Total CAD, the Veribest Electronics Database, and Simulink (Functional Control Systems) Computer Aided Design Engineering package, helping the disabled, and motorsport.

Achievements

Engineering is a highly interactive discipline – as a Specialist, I continuously learnt from the tip of the skills iceberg which is within me and my colleagues across Europe, to design and prove a highly complex Panavia Tornado engine. I read and correspond on many things to MP, Lord and Government, and take part in many talks in Engineering and Physics.

- I created the first software simulations in 1968 for the three shaft RB199 engine for the Tornado aircraft, and thence learnt about how the engine behaved before it first flew.

Supported British Aerospace (BAe, later BAE Systems) with their concept for the aircraft using my simulations for the engine concept before it was designed.

- I created engine specific telemetry analysis routines for British Aerospace at Warton, to improve advice to the telemetry room crew, and thus to MRCA Test Pilots whilst still in flight.

- I proposed solutions for Tornado engine operations problems for the RAF and its partners. I maintained contact with the RAF Engineering HQ.

- I created the Tornado Engine defect acceptance and repair advice network within Rolls Royce plc, and coordinated with our partners in Europe, for repair of RAF components.

- I Project Managed Rolls Royce's and Europe's first successful Single Crystal Turbine Blade design and make programme (created and presented paper in 1987 in Tokyo).

Visited Riyadh to present proposed modifications to combat gypsum powder in the lower atmosphere for RSAF Tornado (Royal Saudi Arabian Force).

- I confirmed that Weibull statistical analysis techniques most reasonably represented the observed RB199 component reliability environment in Service. Saved grounding aircraft on at least one occasion.

- I received a congratulations letter from the RAF in 1984 addressed to Director Bristol.

- I managed a 6 month design programme for an electronic test box for the US Navy Goshawk, achieving on time.

- I launched the inaugural HISSC (High Integrity Systems and Software Centre) in Rolls Royce plc.

- I proposed a Safety Strategy for Britain's Railway Network to the Deputy Prime Minister following the Ladbroke Grove crash in 1999 – fundamentally, we need the coloured external traffic light signals to be repeated on a miniature display in the Drivers Cab.

- As a RemapGlos Engineer collaborating with others, I work for free producing custom, one-off equipment for the disabled and we do not charge the client. As one time Publicity Officer I have produced several local Yearbooks two yearly, and have completed over 130 jobs. I created our second website, now superseded by a third website at www.remapglos.org.uk. I self generated a Remap presentation package for various charity awareness seminars, and thence received more jobs.

- At one time I was a volunteer carer and trustee with the Headway charity in Gloucester. I used to assist attendees in reading, writing and numeracy, and constructed tasks for them to do related to normal living and helped with the use of money, and how to deal with bus & train timetables. As a Trustee, I initiated fund raising activity successfully (the local Rotary for example) with self-generated presentation.

- A successful experienced motor sport competitor and organiser, 1980 – 1993. I was the Company Secretary of Tavern Motor Club Ltd in 1980-82, awarded the Committee Rose bowl in 1981, Chairman in 1982-84, and Clubman of the Year in 1986. I was Clerk of Course on 11 occasions, and Results Chief on 8 occasions, having been an organiser on a total of 24 occasions. I received a congratulations letter from The Avon and Somerset Chief Constable in 1983 for the good organisation of the October 1983 Orion Road Rally across the Brendon Hills in Somerset.

CHAPTER 1

EDUCATION AND INHERITANCE

I went to School from when I was 4 years old, and finished up at Erith Technical School until I was 17 in 1960. I was at a Convent School in Erith until 7 years of age, then Crayford Junior until 10, then Welling Secondary Modern until 12, then Erith Technical School where I was the first to take A levels in that school. I was thus educated in Roman Catholic schooling until 12 years old. I did not pass the 11 plus exam, perhaps because we were not allowed to practice in those days and it takes me a little time to adjust to a new situation, and I did pass the 13 plus, but not at high enough level to go to Grammar school thank goodness, for I would have hated the speedy cramming you get there. Welling was a very good school and I thoroughly enjoyed music, geography and algebra there. I still like to observe and think about things before launching into loud responses.

But at 4 years old I thought I had completed my education, for at lunchtime I returned home on the bus, all on my own, with no money (the bus conductor allowed me to travel for free), for I had done school! My Mother was astonished that I had done that all on my own at that young age. I was 7 years old when my 5 year old brother, Patric, asked my Mum where babies came from, to which I apparently promptly replied "From the church silly". I broke my left arm twice before I was

5 years old, both times at the elbow and one of those times I was told I fell down the concrete steps at our back door. I remember the steps, but not the fall, because we did not leave that house In Barnehurst until I was 6 years old. I also had a hernia operation at about that age (I was embarrassed at having to stand naked in front of a group of students while they looked at my dropped genitals), and I remember that. I also remember having a tonsillectomy and coming round with blood all over the place when I was 6 or 7 in Charing Cross Hospital in London, which ice cream helped! My mother also made sure I empathised with the disabled, and I was brought up to be honest and never ever to lie or mislead.

Mum told me after a parent-teachers evening that I rarely volunteered information in response to a teachers question in class, but that I always seemed to know the answer when directly asked! When I was 9 years old I used to cycle a couple of miles to school and raced with my friends – unfortunately my legs had to fly with the pedals to keep up with them because they had gears and I did not, but I kept up with them, breathless! At the same age, I had my first and only fist-fight – the boy next door, Chooky, had threatened my brother for some reason and I did not like what he did, so I forced him to accept that he had lost! I have suffered the cane on my hands which left weal's at Junior School when the paper was so porous that my ink spread sideways from the character making it very fuzzy (not my fault at all, but the silly teacher caned me because of the paper!).

I was 11 years old in my Dad's photo above. I was a Scout from 7 years old (in the Cubs) until I was 17 when I had been in the Guy Gibson patrol for many years in the 9th Dartford Scout Group. I finished up with a Queens Scout badge, and eventually became a Venture Scout with a Bushman's Thong. I had badges for astronomy, forestry, despatch rider, rescuer, public health, and ambulance, and the 1st Class badge covering things like first aid, tree felling, swimming, mapping, compass, highway code, 10 nights camping, and so on. I went to the Scouts Gang Show in London with my Father several times – thoroughly enjoyable singing and stage work. He allowed me to drive his coach/bus around a private car park when I was 12! I mostly used to eat bread and jam and corned beef hash (potatoes, greens and corned beef).

I am now glad that I had a Technical School education, for I was better able to grasp the logic of the teaching. I did not like History for it was all dates and no logic, and I had no interest in Kings killing each other. There seemed to be little debate going on in those olden days, and arguments were won because of aggressive personal need resulting in physical strength competition not intelligent logic – daft or what! Geography became all about Gross Domestic Product and its constituent parts and at that time also of little interest to me logically. There was no biology course, but I find myself quite interested in evolution and neurophysiology now, which I started when I was in my 50's: see ISBN 978-0-19-923085-3 "Why Evolution is True", and ISBN 0-340-50634-2 "Neurophysiology".

I liked school learning about music and the notation used when I was 12. I spent many an hour as a teenager listening to my Dad's classical 33rpm records (Tchaikovsky's Romeo & Juliet and Rossini's William Tell Overture I well remember) which I took great care over, for the tracks could so easily be damaged by my fingers let alone needle skid – and I was listening when my parents were out at work and did not know! I have enjoyed classical music ever since and attended concerts, and I am certainly attracted to Jean Michel Jarre, Phil Collins, Dire Straits, Queen and especially Michael Jackson too. I particularly like Michael Jackson's Black or White video on YouTube (the official version), particularly because of the many glorious and happy faces he shows at the end!:-

I gained 7 O levels in maths, physics, mechanics, metalwork, technical drawing, English and additional maths (85% for that! Which meant I had mastered the easy stages of differential and integral calculus by the time I was 16). I gained 3 A levels when I was 17: pure maths, applied maths, and physics. I received the 1959 one mile walk medal at the Kent County AAA Junior championships when I was 16 years old. I went to the RAF Biggin Hill Airshow when I was 17, and watched a Meteor land when its left main undercarriage collapsed, causing it to scrape its wing tip along the tarmac. Then a Spitfire flew past right to left and I was leaning over the barrier looking left to watch it, when with a surprise whoosh, a fast Hawker Hunter followed it immediately afterwards – I was thus firmly hooked on aviation!

I spent time with my Dad helping to repair engines and dynamos, and I eventually was taken on as a Student Apprentice, but I originally applied to and was interviewed by Bristol Aircraft Ltd. At the interview I said "I was fascinated by the cams that made the valves move on my motorcycle engine, so I want to work on engines please", and they said "that is alright, we are interviewing for both Companies", and they took me on as an Apprentice to Bristol Siddeley Engines Ltd at the local Bristol Aeroplane Company Apprentice College (now Filton College, but where have all the previous machine and workshops gone)! I moved to Bristol from Erith to do that, and stayed with my Gran and Aunt Tris who lived in Keynsham and then Bristol for all of those apprenticeship years.

I was a Student Apprentice from 1960, and I finished up with a Bsc(Hons) Upper Second, coming top of the class on Fluid Control Systems. The list of exams I took in 1965 were: Strength of Materials, Aircraft Structures, Mechanics of Machines, Aerodynamics, Thermodynamics, Applied Thermodynamics, Analytical Methods, Mechanics of Fluids, Aircraft Engineering, Aero-Engine Design, Theory of Structures, Fluid Control Systems, Fluid Control Systems and Machines, and Industrial Administration. My final year project with Rod Clifford was "Supersonic boundary layer separation pressure gradients", completed in May 1965. The separation was caused by a small step upwards and we measured many pressure tappings upstream of that step – this was triggered by Concorde thinking - like how proud could rivets be allowed on it.

I was a contender for Vice President of the Students Union in 1964, and I took part in the hustings on the main campus and on the remote one where the architects were taught. The vote went my way, but of course the architects voted en bloc for their own man! I was thus Vice President in 1964/5, and was told by my tutor that was why I was unlikely to obtain a First! I enjoyed much science fiction, Arthur C Clarke and Isaac Asimov I remember most, but I also read Analog magazine sometimes. I also worked with Rod on a Rag Stunt in putting a rowing boat on top of the four storey radar building opposite the University!

– photo in the Bristol Evening Post. I won prizes over several years for being best Student Apprentice, but I also did well playing bridge with Matt Westley (an electronics student at what became Bath University), Ian Shepherd (pharmacy), and Dick Toft (chemistry) during our lunch breaks.

During my Apprenticeship, I spent a few weeks in the Whitchurch Hercules Engine repair department, and stripped Hercules engines as part of my job to learn, with help from the permanent craftsmen of the day. One engine I helped strip, unusually had an upturned piston as a result of a broken conrod! I was given a cash bonus by the craftsman I was working with – he obviously appreciated my help with his bonus that week! I also worked in the Design Drawing Office on the Palouste engine, and the Stress Office, as well as the Jig and Tool Drawing Office which faced the Filton runway. From there I saw several Vulcans scramble for take-off one day (it must have been 1963 then), and they seemed to drift partly sideways around the right hand corner at the end of the runway, before taking off with a very loud bellowing noise and lots of smoke! I joined Bristol Siddeley's Advanced Propulsion Research Group (APRG) in 1964 in my final apprenticeship year because I was good at that sort of work (and I was wanted by John Lane the Performance Department Manager - part of APRG), and I worked on Sea Dart double cone intake shock geometry (and later I worked on the Odin combustor efficiency correlation around the whole flight envelope). I was selected to make a performance assessment of a Russian air-breathing missile photograph in flight – during the Cold War, and I kept very quiet about it, to best protect the photographer!

While there, I spent a short time on BS1012 hypersonic (Mach 5 +) scramjet engines and once experienced the actual use of a hypersonic shock tube. I learnt that running lines on compressors were just being presumed and not computed – how were those compressor running lines controlled I asked myself? By slide rule calculation and with impossible difficulty because we were unable to properly calculate the interactions of compressor and turbine components together. The complexity of the calculation meant that it took a long time to even calculate just one flight point given the presumed assumptions, and it was impossible to calculate where the running lines actually were. Concorde's and TSR2's two shaft Olympus engines and Harrier's Pegasus engine were all slide rule calculated too, and therefore required extensive testing to determine their running lines relative to their compressor surge lines, which I will explain later.

MARRIED

I married Mary in 1966 (she died in 2009 of lung cancer unfortunately) as a result of meeting her during her A level Pure Maths studies in Bristol Siddeley in 1964, and at the local Blue Notes Jazz Club sessions in the Llandogger Trow Inn in Bristol. Mary used to sit against the back of the piano enjoying Roger Bennetts sax, also thoroughly enjoying the drum solos, as I did. I used to organise the bi-annual dance sessions in the College Hall, and we always enjoyed Jiving to the Blue Notes there too, let alone Line Bop and other energetic dances. We learnt the Waltz and the Quickstep then too. We had our children a year later and nearly three years later, which caused some significant changes in our life style! My second was a very heavy baby at 9lb12oz which was quite exhausting and hurtful for Mary, and probably him too, but Mary's waters broke two weeks earlier than he was born, and he suffered Bronchitis for the next 7 years as a result I think. Mary loved France and especially the huge fields of bright sunflowers.

I joined the local Gilbert and Sullivan Bristol Catholic Players between 1969 and 1976, because they were well known as the best and good amateur G&S artists, led by a Monsignor Joseph Buckley. He was very good at the choreography arranging to use all of Gilbert's original stage craft notes in the production. Donald Adams was a good friend of his and he always came to see the Players annual production, chatting to us all in turn as celebrations passed after the event he attended. I mostly constructed the scenery and stage managed several productions with my good friends Trevor Morgan and Frank Shipsides, the celebrated Bristol painter who designed the sets. Trevor flew Mosquito's in the second world war, but he suffered heart failure and died, so I took over his job as Scenery Manager in 1977, but not Stage Manager that year because I was a Gentleman in the Mikado instead.

I took a part in a couple of other productions: Hadrian VII as a clergyman, Pirates of Penzance as a pirate in the chorus, and in 1976 in Ruddigore I played Old Adam Goodheart a faithful servant to Sir Ruthven Murgatroyd. Mary, my wife, played a non chorus part too, as Ruth, a professional Bridesmaid! She also was assistant wardrobe mistress to Kay and they made all the costumes for the cast for several years. I left the Bristol Catholic Players when I went hang gliding in late 1977, and stopped going to church and stopped drinking alcohol of any sort subsequently, having enjoyed my University and Players years on beer!

GENEALOGY

Attached is the data that I have now found about my predecessors in order from bottom to top of the page. Those nearer the top of the page are the oldest, and were very difficult to trace for the Gloucester records were so sparse, and there are blanks because I have no data. Also, for example, the Quinn dynasty was obviously alive back in the 1700's in Ireland, so it is just that I have not found one in my line before 1847. I went to Dublin in 1995 to try and find some background to Quinn, but many records were destroyed in the 1846 Famine and I found nothing that helped. I now know I should have checked Northern Ireland's records. And the same lack of genealogical evidence goes for Leopard before 1818, and so on. I also went to Roscommon in Eire to see the Famine Museum there and was horrified by what the British

Government had done during the famine - they only gave money to the Irish if they sold the food they needed to avoid starvation!! And then the UK Parliament complained about the money required!! No wonder the Irish wanted Independence – and got it.

Note that I have many birth and marriage certificates, all indicated by the * alongside their marriage (m*) or birth beside their age at marriage (like *23, tailor) also indicating their profession. Oh, I have the names of some of the children. Joseph Charles Leopard (my maternal great-grandfather) and Eleanor Ann Crowe (my maternal great-grandmother) had 7 children as follows: Trubshaw Thomas, Sydney Church, Gertrude Eleanor, Joseph Ezekiel, Garnet Evelyn, Eric Andrew (my grandfather), and Harold Albert in order of birth. Then Eric Andrew Leopard (my maternal grandfather) and Emily Muriel Leopard (nee Workman – my Gran) had Eric John, Winifred Mary, Peter Garnet and Patricia Gertrude (Tris). Interesting names! Also James Francis Quinn (my paternal grandfather) and Alice Wallace (my paternal grandmother) had children called: Phyllis, Francis, Lally, Una, Gerald, Desmond, and Bernard Eric.

A Summary Genealogy

G^8 grandparents

Thomas Bartlett + Sarah Davis	m.25sep1705 Bisley

G^7 grandparents

Thomas Bartlett + Sarah (died 28jan1761 age 44)	
Thomas or Thomas or John Collins (Mary bap 2apr1735 or 5feb1746 or 21dec1749)	Mary bap Bisley
John Mayor (Lydia bap 7aug1743) NB John could be one of four married between 1727 and 1734 in Bisley	

G^6 grandparents

Thomas Bartlett (24, broadweaver) + Lydia Mayor	m.4aug1764 Bisley
Mary Collins (Elizabeth bap 13sep1768)	Bisley

G^5 grandparents

Thomas Bartlett (26) + Elizabeth Collins (22)	m.5march1791 Bisley
Henry or Robert Gardiner (Anne bap 29may1791 age 2 – or 31oct1790)	

G^4 grandparents

Robert Bartlett (20, weaver) + Anne Gardiner	m.31march1812, Bisley
Henry (a smith) + Sophia Richardson (widow age 68 in 1851 census, a weaver)	Stroud Bourne
Ann Workman baptised Joseph Workman (alias Mason?) on 10oct1810	Bisley

G^3 grandparents

William + Nancy Leopard (labourer)	m. before 12jun1818
William Wake (tailor) + MB (Mary Anne born aug1823)	m.9nov1823 Buttsbury Essex
Peter Crowe (farmer)	m. before 8jan1828
James Church (blacksmith)	m.before 8jan1824
Joseph Workman (25, widower, baker) + Anne Bartlett (22, they had ten children)	m.28oct1835 Bisley
Amos Richardson (18,blacksmith) + Mary Ann Mason (19, servant)	m*.3jun1839 Stroud

G^2 (meaning great great) grandparents

Francis Quinn from Northern Ireland + Ms Rice	m.before 22nov1847
James Murray (bricklayer)	m.before 22nov1855
... Rule	m.before 1jun1858
William Leopard (25, railway timekeeper) + Mary Ann Wake (20)	m*.12jun1843 Ingatestone
Ezekiel Crowe (<21, stoker) + Eleanor Church (21+)	m*.8jan1844 Stoke Ash
John George Workman (*18, baker) + Emily Richardson (*20,cloth-maker) (5 children)	m*.25dec1860 Chalford
John George Bennett (wool sorter) + Emma Aldridge	m.1862? Minchinhampton

Great grandparents

Peter Quinn (24, hatter) + Mary Jane Murrey (16, paper sorter)	m*.22nov1871 Norwich
John James Rule (cork/stencil plate cutter) + Alice Murray	m.before 1 jun1874?
Joseph Charles Leopard (*21, fitter) + Eleanor Ann Crowe (*21) (7 children)	m*.27mar1869 Woolwich
John George Workman (*25, provisioner) + Mary Jane Bennett (*23,clerk) (4 children)	m*.15jan1887 Sydney

Grandparents

James Francis Quinn (*23,tailor) + Alice Wallace Rule (*22) (7 children)	m*.3aug1896 Stepney
Eric Andrew Leopard (*30,electrical engineer) + Emily Muriel Workman (*27) (4 children)	m*.27nov1915 Deptford

Parents

Bernard Eric Quinn (*27 engineering inspector) + Winifred Mary Leopard (*20 secretary)(3 children) m*22feb1941 Dartford

Note that if you displayed the Royal Family in this way, you would find that Prince Charles and Princess Anne have none of Prince Albert's or Queen Victoria's X or Y chromosomes!

A designer and artist of many oil paintings, my uncle John Leopard wrote ISBN 978-0713721386 "Ships in Bottles: A Modeller's Guide" published in 1990. He also landed on D-Day very early in the morning on Juno Beach with the Canadians in his modified Sherman tank – he survived the war.

ASTRONOMY

I visited the Crayford (Kent) Astronomy Group when I was about ten years old to view Saturn through their 6 inch (150 mm) reflector telescope, and this really turned me onto astronomy. I still take much notice of astronomical findings about other galaxies, stars, and their planets, and look forward to the regular magazine that I receive from my membership of the USA's Planetary Society, who I have found to be useful advocates to the USA President and Congress on NASA support – I do not want to send humans to Mars before about 2040, but I do support establishing a base on the Moon, which I consider is an essential prerequisite for humans visiting Mars.

I will be about 90 years old by the next time we visit Uranus or Neptune, and I then also look forward to exploring Jupiter's ocean-bearing moon Europa and Saturn's sixth largest moon, Enceladus, which has jets blowing from the poles - an instrument on board the Cassini spacecraft detected the presence there of a biomarker called "molecular hydrogen"!

I also want to visit Saturn's Titan again. These are science missions which I see as higher funding priority than humans returning to the Moon, so let us not lose these relatively low cost missions in the rush to the expensive Moon please.

I have also donated to the Laser Bees Near Earth Asteroid deflector described at http://www.planetary.org/blogs/bruce-betts/20131029-laser-bees-update.html

The Earth is four times the diameter of the Moon – have you ever wondered what the Apollo astronauts saw from the Moon? A very large Earth! Apollo's 8 and 13 never landed of course, but both went around the Moon. The maximum speed of the Apollo spacecraft was about 25,000 mph and slowed down as it travelled further from the Earth, while the International Space Station (ISS) travels at about 17,500 mph continually around the Earth. To get up there it had to fly at Mach 27 through 200,000ft altitude, where it was 300knots IAS. When I look at the ISS at dusk flying overhead, I astonish myself since 10 minutes ago it was over the USA, and that in 30 seconds time it will have flown from Cardiff to London! I have written to my local council in the past about light pollution – I wanted the street lights turned off at 1am and back on again at 5am, so that we astronomical viewers living in towns can see more of the stars please.

The Sun is due to become a red giant star in four billion years time, and it will consume the Earth and the Moon then, but the Apollo vehicles will still exist on the Moon's surface for erosion is virtually zero there. Not like our 5000 year old Pyramids and Stonehenge which will be subject to considerable erosion over that extended period of time.

I constructed this picture from my own observations of where the Apollo vehicles landed.

BRAIN TYPES

There are at least eight types of human brain – after discussion with Vera in her 80's (who was once Headmistress at a Junior School) I think these are:

Artiste (portrait painters and sculptors for example),
Kinetic (ballet dancers and gymnasts),
Numbers (mental arithmetic),
Maths (calculus logic),
Words (Shakespeare),
Crosswords (cryptic clue manipulation),
Diagrams (image recognition)
Music (composer or orchestra player)

And there are all grades between expert and not so good. I am good at numbers, maths, diagrams, and kinetic (as a motor-sport driver), but not so good at artiste nor words (as images, my word spelling and sentence construction is good, but I hate Shakespeare's words for they are so many and are difficult to understand). I use simple words that virtually everybody can understand, but I cannot do crosswords for I am unable to recall words easily, and do not like manipulating letters in anagrams, and I am unable to follow cryptic clues, but I enjoy Sudoku. I once had a competition with my friend Vera which she won easily. How many words beginning with the letter S can you write down in 60 seconds? I managed 7 but Vera managed 25, so she was obviously a wordy person ten years older than me in her 80's. So age does not matter, but brain type does. Both Vera and my wife Mary enjoyed crosswords, Mary particularly liking the Daily Telegraph crossword, but not The Times for she said that she better understood the Telegraph creator's mind - I could not do any of them at all! I enjoy music of many sorts, but I am not a composer, and I admire the orchestral violin and piano soloists for their exceptionally good memories. Kids all learn differently, using different techniques to learn.

I went to the Stratford on Avon Royal Shakespeare Company's production of Romeo and Juliet a few years ago, and tried to follow it. Within half an hour my brain had given up – my concentration level to understand all of those very rapid words gave me a headache, so I left before the end. However, the noise on stage was very aggressive for nearly all of that time – demonstrating that R&J was not a love story at all, but a HATE story where the warring factions kept having a go at each other. A kinetic presentation!

LOGIC and MEMORY share the same Brain space

The next two "charts" show MEMORY and LOGIC, and the illustration of the wrong answers to the sums is meant to show how people around the world might see themselves. Note that our brains have hard wired instructions genetically from birth which demand that we eat and keep our heart beating, and software that learning stores in memory to guide us as we grow older. I liken this to the modern computer where ROM hardware starts the computer, and RAM software is what we buy or use or delete or replace to logically produce the result we want, that we also store in memory as non logical memory. The more memory we consume, the less we have available for software logic.

Clearly, if you have no logic or if you do not use it to think about the rest of the world, you never change. Imagine living in a country where Hinduism is the dominant religion – it is probably OK as a religion, but thinks it is the only one that is right, and I therefore assign the sum 1.1 + 1.0 = 2.3 to it – not that the sum is what they actually compute, but it is a representation of what they think about life and their religion is not the same as every other one of course. Similarly the Roman Catholic Church thinks it is right, so I assign 1.1 + 1.0 = 2.2 to that. Then 1.1 + 1.0 = 2.1 is Humanism, and 1.1 + 1.0 = 2.0 is the Church of England, and 1.1 + 1.0 = 1.9 is Islam. Thus my scattered MEMORY illustration below, shows that none of them think closely together - it is an illustration of difference.

While my LOGIC chart below shows that even though they are scattered apart by distances, they all think the same accurate logic, and the future is thus protected, and extrapolation is easy and possible.

MEMORY

1.1 + 1.0 = 2.3

1.1 + 1.0 = 1.9

1.1 + 1.0 = 2.1

1.1 + 1.0 = 2.2

1.1 + 1.0 = 2.0

If you only have memory, and no logic......no change, future = past, doomed forever, and you will never be able to extrapolate – try 1.3 + 1.4, and not be able to answer it, for you have no logic. And thus the memory-dictator no-logic Pope imprisoned logical Galileo in 1563 for proving that the Earth was not the centre of the Universe, and that the Earth went around the Sun.

LOGIC

1.1 + 1.0 = 2.1

1.3 + 1.4 = 2.7

0.5 + 1.0 = 1.5

10.9 + 2.1 = 13.0

0.1 + 19.0 = 19.1

Here a system is evident..... this is OK now: future thinks

Our brains, although huge billions of neurons, have a limited capacity. We have to stop teaching young children to remember such memory intense detail as religion imposes. We must not expect them to remember by rote the Bible (the Dark Ages resulted) or Shakespeare (far too many words for youngsters to understand), for that occupies so much memory that logical thinking becomes impossible. The only rote memory that should be imposed is all twelve times tables, and basic and simple language. Logic uses memory to store the process or procedure, while progress is its logical questioning and thinking during iterations that needs intermediate storage in memory.

CHAPTER 2

MRCA-TORNADO CONCEPT EVOLUTION

SIMULATION

I worked with SNECMA (Melun, Paris) in 1966 on the M45G engine for the AFVG (Anglo French Variable Geometry) aircraft, then I created the Bristol Engines 1967 performance software simulation in Scheme P for the two shaft BS143, and the same in 1968 for the three shaft RB199, the MRCA engine. The MRCA was the name of the aircraft before it was called Tornado. Later, in 1971, I added the extra part of the simulation for the transients of acceleration and deceleration in the non-reheated 3 shaft engine range. The reason I was creating the computer simulations before anybody else on the MRCA engine in Bristol, was because I had been so interested in how the performance of an engine was worked out by slide rule, and how to determine where the compressor running lines were, before we had the computer to perform the simulations properly, and I was very keen to create, for the first proper engine simulation ever, the software on the new KDF9 computer facility – our old DEUCE computer could not do it.

I went to SNECMA with Jim Harris and Brian Yates, to help prepare the brochure in Monsieur Habrard's Department for the M45G, and spent three weeks there plotting computer output onto graphs of non-dimensionals, as well as eating raw meat for lunch (as the French liked it back then, but I did not!). I stayed at the Hotel du Grand Monarque in Melun, and still have a few letters that Mary wrote to me there, before we were married some 6 months later in August 1966. I also still have a painting of me that an artist made in Paris in 1966 – very blue eyes!

The graphs were to be supplied to the Airframers and the Ministries to allow them to use them wherever they liked around the flight envelope, but they gave no digital figures for thrust, fuel consumption or air flow at any specific flight point and engine spool speed – these all had to be deduced from the graphs, by substituting temperature and pressure by hand to unravel the non-dimensional into a digital dimensioned figure. It was because of this non-dimensional experience, after the AFVG was cancelled, that we in Bristol decided we would produce brochures with digital outputs tabulated all around the flight envelope so that the engine data could be immediately seen without further manipulation. This we did by asking our Computer Department to tabulate the simulation outputs for us from Scheme P and in our requested format and tabular pattern.

Incidentally, my first flight ever was to SNECMA for the first time, on the Nord Noratlas to Paris from Heathrow. Previously, I had travelled on the ship ferry to Cologne with the Boy Scouts in 1955 and then with my parents later. After a couple of trips to SNECMA, the Noratlas propeller reversing mechanism failed and we were subject to very heavy vibration during landing – this jammed the access door and we were thus marooned on the apron for a while, until somebody sorted it out!

On the MRCA engine we used a new, and for the first time, computerised software simulation on Bristol Siddeley's new KDF9 computer, which I created on behalf of the whole of Rolls Royce firstly for the mixed 2 shaft reheated engine, then for a mixed 3 shaft reheated engine, which was for the first time in Rolls Royce (it turned out that the Derby work in 1968 was for a small civil non-reheated unmixed engine type, not at all representative of a military reheated mixed engine type). The civil type was unmixed for the bypass and turbine exhaust streams exit their own separate nozzles, whereas the military type mixes the two streams together before exiting the final single nozzle, and if required, burns reheat before exiting that single nozzle. They thus perform differently. One unique and special thing about the BS143 and RB199 was the very short tailpipe, so reheat was burned in the bypass and turbine exit streams separately, mixing them downstream of the flame stabilisers as the nozzle approached. This also was Bristol's idea and quite different from Derby experience on their Phantom Spey engine where the mix occurred before burning reheat, and which therefore had a very long and very heavy tailpipe.

Reheat increases engine thrust, because the gas expands because of its high temperature, and thus allows the Control System to open the nozzle without changing the compressor operating conditions. This provides a bigger area and higher velocity at the nozzle exit increasing thrust, and at the same time reduces the annulus area and thus the aircraft back-end drag of the nozzle flaps themselves.

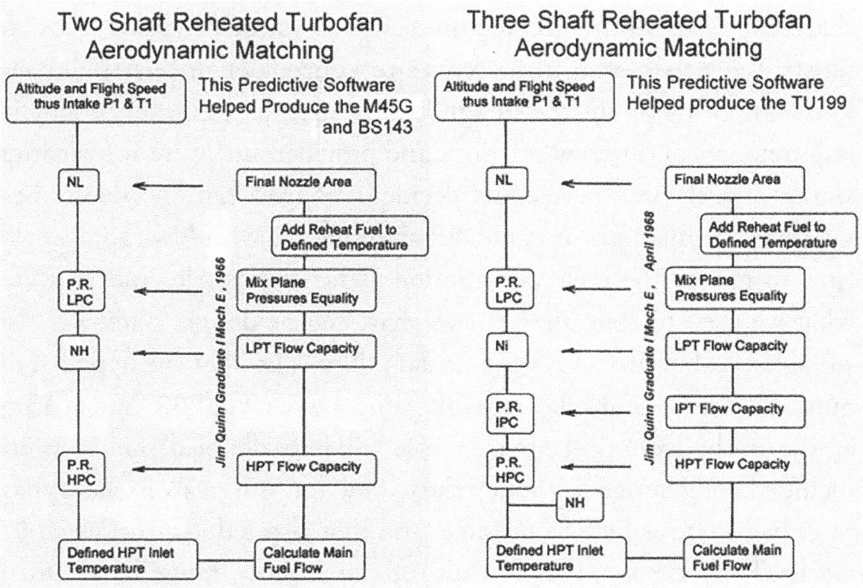

I was a member of the RB199 Performance Department who were all able to use my simulation software to provide detailed numbers internal to the engine, which design, stress, installation, controls and aerodynamic specialists needed to do their jobs. My software also produced the installed thrust, fuel consumption and air mass flow numbers that an aircraft company would need. The above diagrams shows the logic of these two and three shaft simulations, where the left hand column of each is "predict as best you can", and the right hand side close loops (matches) by adjusting those predictions to improve them to an accuracy of 0.1% of the right hand side fixed values. The turbine flow capacity was calculated from the predictions information, by assuming that the throat was choked – that is, the flow was Mach 1 at the throat. Note that I have simplified the diagrams to show the biggest influences – no air cooling of the turbines for example, but which was actually in the simulation. Hilary Castle was a young graduate who came along to help a few years later, and she used to update the software simulations too.

In the concept stage we had used my simulation to generate the two shaft engine performance requirements at various bypass ratios, to satisfy the needs of British Aerospace concept engineers. I did not know anything about the design of the aircraft, but understood the requirements of different missions and provided sufficient information so that British Aerospace could decide the bypass ratio needed to best satisfy those missions. It came out at about 1.2 with low fan air mass flow to reduce the fuel consumption as far as possible, and that was what we used to generate our two shaft engine design. This was also suitable for the three shaft engine data. Note also that the bypass ratio quoted is the standard day take-off value, but that at 800 knots IAS or at idle on the ground, the bypass ratio will have doubled. The missions included long range without reheat and for this reason the bypass ratio had to be as high as possible, and that also reduces the Mach 0.9 sea level dry thrust. This low altitude maximum range requirement also meant that the engine would have to be operated at maximum continuous dry thrust for a long period of time, and we did not want that thrust to be too high because fuel consumption would thereby be increased, so that was how the mission optimisation was performed. Maximum reheated thrust was required at Mach 0.9 low altitude to sustain maximum aircraft G in a turning combat manoeuvre, and to provide maximum acceleration at high Mach number at altitude too, and these were increased by the higher bypass ratio. Because of the high sustained Flight Mach Number, Stress Office proposed a limiting HPCompressor delivery temperature line on a graph against NH (HP Spool Speed), and I put this into the simulation too, so that we could determine the control function needed in the MECU (the engine electronics).

The simulation used enthalpy, h, and entropy, s, as functions of temperature, not the specific heats at constant pressure, Cp, and constant volume, Cv, for those have to be averaged over a temperature change, and we wanted to be precise. Enthalpy is merely the product of temperature and Cp at that temperature. Work done, for example, is the difference between the enthalpies between two temperatures, and pressure ratio is the difference between two entropies at their two temperatures – our definition of entropy. A compressor demands energy to work at all, and the turbine provides that energy. The HPTurbine drives the HPCompressor and the aircraft gearbox, the IPTurbine drives the IPCompressor and the LPTurbine drives the Fan (the LPCompressor). So, given the compressor inlet temperature and pressure ratio and an isentropic efficiency to define the temperature at exit, the energy required comes from the enthalpy equation. Given the two temperatures, the enthalpy change is the work required for unit mass flow, so they are multiplied together to provide the over all kW (kiloWatts) needed to drive the compressor. The turbine calculation then used the inlet temperature, the kW, and the mass flow in the turbine to calculate the enthalpy change required, from which exit temperature comes out, and pressure ratio from the efficiency and entropy change at the two temperatures.

Condi Nozzle Studies

In 1968 we studied a condi (convergent-divergent) nozzle for MRCA, but decided that, for our missions, it would not be sufficiently beneficial to warrant its extra mass relative to the convergent nozzle we have today. However, it appears to fit Typhoon's engine and role better for they have a condi nozzle.

The "ConDi Variables" diagram below shows such a nozzle, with its variable parameters above it. You will see that the entrance to the nozzle is at subsonic flow and large area, then it passes through a throat at Mach 1.0, and into the increasing area divergent section where Mach Number, M, is supersonic.

"a" is the speed of sound which varies with static temperature t, and V is the absolute velocity of the gas along the axis of the nozzle. p is the static pressure dropping all the time to a value above the ambient atmospheric pressure outside the nozzle.

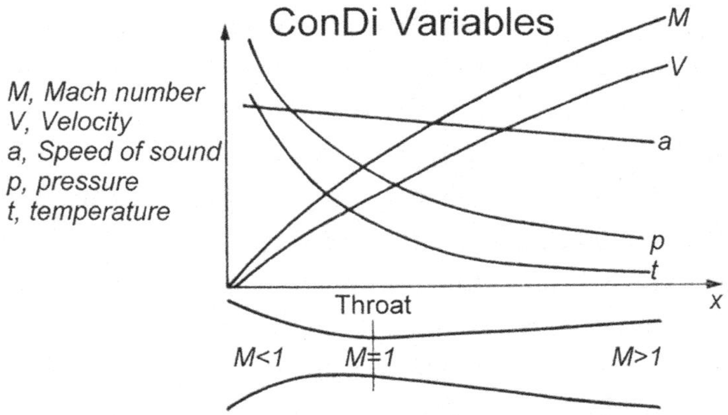

The Mach Number at exit of a condi nozzle is calculated from equations [c] and [d] below, where P is the stagnation pressure (assuming no losses along the condi nozzle) inside the engine, and p_e is the measured exit static pressure, which is not necessarily $p\alpha$. Note that ρ is density andγ is the ratio of the specific heats Cp/Cv. γ for dry air at normal temperature, 288.15K, and pressure, 101.325 kPa, is 1.40, but for a reheated engine exhaust it could be as low as 1.20. Note that Net Thrust (equation [f] below) is calculated from nozzle exit subscript $_e$ and aircraft freestream subscript $_0$. I learnt how to generate these equations during my apprenticeship.

Given the supersonic exit Mach Number, M_2, the area ratio (throat area, A_1, relative to the larger exit area, A_2) can be calculated from equation [e] with M_1 = 1.0 at the throat, which is choked. At idle the engine pressure will be too low for the nozzle to be choked (therefore throat Mach < 1.0), in which case equation [e] can still be used for subsonic Mach numbers too. The reason we use these equations is because we are doing a concept calculation and have not actually made a condi nozzle yet. So we know what p_e we want (and we might try a few different values so that we can draw each result and thence estimate the weight of each, until we are happy), from which comes the Mach Number, and then the actual area ratio. Now we can manufacture one – and test it of course!

Table 1: Variables for a frictionless subsonic or supersonic duct

$$[a]\, Mass\ Flow, \dot{m} = \rho A V$$

$$[b]\, Speed\ of\ sound, a = \sqrt{\gamma R t} = \sqrt{\gamma p / \rho}$$
$$where\ R = 287.058\, J\, kg^{-1} K^{-1}$$

$$[c]\, \frac{T}{t} = 1 + \frac{\gamma - 1}{2} M^2$$

$$[d]\, \frac{P}{p} = \left[\frac{T}{t}\right]^{\frac{\gamma}{\gamma - 1}}$$

$$[e]\, \frac{A_1}{A_2} = \frac{M_2}{M_1}\left[\frac{2 + (\gamma - 1) M_1^2}{2 + (\gamma - 1) M_2^2}\right]^{\frac{\gamma + 1}{2(\gamma - 1)}}$$

$$[f]\, Net\ Thrust = (\dot{m}_e V_e - \dot{m}_0 V_0) + (p_e - p_a) A_e$$

There is another constraint on condi nozzle performance and that is when the engine is throttled back, but well above idle and when the nozzle is still choked (throat Mach = 1.0). At this throttled back condition, the exit pressure, p_e, inside the nozzle then could impossibly drop below the ambient outside pressure $p\alpha$ (as shown in the diagram below), which will result in a normal shock wave part way along the divergent section, which causes the Mach Number to be subsonic thereafter. This allows the pressure to rise in subsonic flow until the ambient outside pressure, $p\alpha$, is reached. However, this shock gives rise to a large pressure loss and the condi nozzle is therefore not recommended for when the engine is throttle back. I tried to construct a software programme to simulate this in 1968 and to find out the resulting thrust and shock position in the duct in area ratio terms – just for fun. I failed to calculate where the shock was because of my inability to properly programme it, but from such things one learns one's limitations.

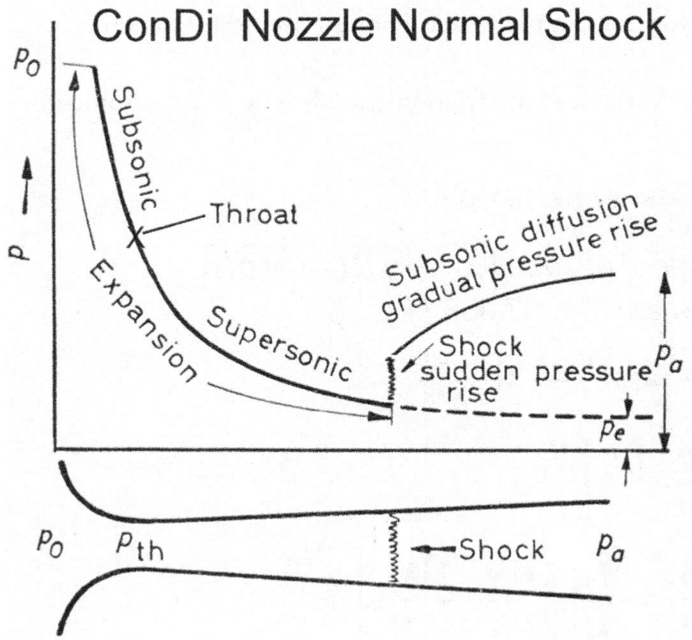

It obviously requires several equations to do it, including the shock loss equations [1], [2], [3] and [4] below, and the effects of sub and supersonic flow from equations [a], [b], [c] and [d] above. The calculation is iterative – in other words predict at what upstream Mach number, $M_{u,}$ the very thin shock will occur, calculate the immediately downstream shock conditions (subscript d), then estimate the nozzle exit Mach number to calculate the area, and change that estimate until the area is correct. Then change the predicted upstream shock Mach number, M_u, until the exit pressure p_e is equal to the ambient pressure pα.

The result is a thrust from equation [f] which is well below that of a convergent only nozzle at that engine spool speed, so you design the nozzle so that it is not divergent at these low engine thrusts – and thus at any condition without reheat.

Table 2: Normal shock equations

$$[1]\, M_d^2 = \frac{1 + \frac{(\gamma - 1)}{2} M_u^2}{\gamma M_u^2 - \frac{\gamma - 1}{2}}$$

$$[2]\, \frac{t_d}{t_u} = \frac{[2\gamma M_u^2 - (\gamma - 1)][2 + (\gamma - 1) M_u^2]}{(\gamma + 1)^2 M_u^2}$$

$$[3]\, \frac{p_d}{p_u} = \frac{2\gamma}{\gamma + 1} M_u^2 - \frac{\gamma - 1}{\gamma + 1}$$

$$[4]\, \frac{\rho_d}{\rho_u} = \frac{V_u}{V_d} = \frac{(\gamma + 1)\frac{M_u^2}{2}}{1 + \frac{\gamma - 1}{2} M_u^2}$$

Increasing area gives a supersonic static pressure fall, while it gives a subsonic static pressure rise, as shown in the condi nozzle diagram above. The extra knowledge required is the constant stagnation temperature T ($T_d = T_u$) and the stagnation pressure P_d after the shock which remains the same stagnation pressure P_d at the nozzle exit (for the flow is normally considered frictionless) so that static pressure p_e can be calculated. The shock stagnation pressure P_d downstream is lower than the stagnation pressure P_u upstream of the shock, and thus a significant loss in stagnation pressure ($P_u / P_d > 1$), but not in static pressure which rises, $p_d > p_u$. The stagnation point is that point on an object placed in the fast flowing gas stream which will actually have the flow stopped (stagnated) – this results in a higher than ambient temperature and pressure, called stagnation temperature and stagnation pressure.

The three shaft engine

Ian Broom, a Bristol Advanced Concepts Design Engineer who had used my data to dimension and draw the proposed BS143 two shaft engine, came with Bristol's Chief Engineer Advanced Projects (Gordon Lewis CBE) and me (the Performance lead on the BS143), to present our work to the Rolls Royce Technical Director, Alan Newton, at Derby in 1968. He said we had done a very good job, and that he had confidence we should now do a three shaft at Bristol! He said that Derby had already told MTU in Munchen that they were going to do a three shaft for that requirement, so the pressure was on Bristol to do it instead. We had done a good job and the good BS143 should have been adopted, but he insisted on a very experimental three shaft at that late time in the MRCA launch programme. He obviously thought that the team at Derby were not able to do it. Bristol Siddeley was one of the largest companies of its kind in the world and offered a wider range of engines than any other manufacturer. They had also offered me a better Student Apprenticeship than Rolls Royce at Derby, and I was not happy that Derby were now dictating to good Bristol Engineering and not used our BS143. Rolls Royce Ltd bought Bristol Siddeley Engines Ltd in 1966 because they were worried about our competitive ability, and unfortunately, Derby were now boss. The bankruptcy occurred in 1971 because they had paid £63 million (a big load of money in those days) for Bristol Siddeley, when the Derby Rolls Royce Board should have had the foresight to make sure their Engineers had enough funds to develop their RB211.

On the RB199 three shaft, it turned out a while later, that we in Bristol discovered Derby had no back ground whatsoever in the three shaft, three stage fan and thus low IPCompressor pressure ratio type of engine – Jim Harris found out about the altitude power offtake surge problem, and I found out about the deceleration surge problem, both caused by the IPCompressor which does not exist on the two shaft BS143 and the surges which would not have occurred on it. I became quite annoyed that Derby had so jeopardised the Defence of Europe and especially the UK! I grew up a few miles from RAF Biggin Hill and under the South East London flight path, so was very much an enthusiast for the RAF, and as a 9 year old youngster wanted to be a Pilot before eventually becoming a Design Engineer. I had nothing to do with Derby for the next 18 years after 1968!

Incidentally, "Surge" is what happens when the airflow in the compressors breaks down because the back pressure becomes too high as a result of too high a fuel flow heat release, normally during rapid acceleration for a 2 or 3 shaft HPCompressor. At higher altitudes we increased the idle speed to reduce the possibility of surge. At a given mass flow in the compressor, the region of normal behaviour is at a lower pressure ratio than the "surge" line – where the airflow angle of incidence on the blading becomes too high and the flow does not follow the blade contours. The result of surge is that the high pressure combustion chamber air "surges" forward out of the aircraft intake in an explosive manner. The intake itself thus has to be built of a sufficiently strong material structure to stop such a high pressure buckling or exploding it. The higher the combustion chamber pressure, the heavier the surge.

See https://www.youtube.com/watch?v=R-UU0pgF5oI for a high spool speed series of surges.

All Derby knew about was the RB211 with its single stage fan and thus high IPCompressor pressure ratio yet still four years before it first flew in 1972, so they had not even got to the point in 1968 where the three shaft idea in any Rolls Royce engine was a success! And they had no altitude power offtake demand whereas MRCA did have very high power offtake wing sweep, and we had a reheat pump too, which took something like two thirds the power of wing sweep at maximum reheat at sea level, so I added that computation as a function of fuel flow into the engine simulations too. We in Bristol did not know this three shaft RB211 “four years yet to come” story at the time. The MRCA programme significantly suffered as a result in its early years, for the change to 3 shaft had occurred late in our programme, while our two shaft BS143 was based upon the 8 years of successful Pegasus engine flying in the Hawker P1127/Kestrel/Harrier which first flew in 1960, so we had plenty of background to justify what we had done. The three shaft concept came along so late that we did not even have the time to be normal and perform the component aerodynamics testing before the first engine ran, whereas it had already been beginning normal plannng for the BS143 concept!

But we did successfully develop the RB199 even though we had a poor, very low thrust, start – Bristol, MTU Munchen, and Fiat Torino eventually made the three shaft RB199 work, and work well, albeit with a higher than predicted fuel consumption because of the need to produce a high surge line (and thus compromise efficiency) on the IPCompressor, to cope with the altitude wing sweep power offtake especially during decelerations. The fuel consumption had also been worsened by a late MoDUK demand for more thrust at Sea Level high speed, in conflict with the need to keep fuel consumption low. However, it was close enough to our concept work that Tornado did exactly what was predicted – a very capable aircraft. The BS143 would have made a better engine, and would have supported the Panavia MRCA development programme much better than the RB199 did, for it was started earlier – procedure is important, and the Technical Director should be made aware of that.

In retrospect, a Technical Director has no authority to change the schedule of work on a Project, merely being employed to consult on the technology level used on that Project. The Rolls Royce Engineering Director (the Technical Director's boss) has Authority over such Projects, and he it was who should have been the Authority here, but he was absent! A Project is about the schedule of works involved throughout the Organisation, not just its technology, which is a fall out from the Programme.

During the workup to the 1969 TS1800 RB199 Specification (which I spent a great deal of time on, with Doug Shoemark and Graham Fellows editing the words) it was agreed that we should use the USA's MilSpec definition for it was to be a NATO Secret engine for Europe – this MilSpec comprised many questions to which we had to supply answers. I am a strong believer in the Supplier writing the detailed Specification, for only that Supplier will know what tools and people will be available during the programme. A Specification written by the Customer will not know any of this, and it will thus be too academic, but the Customer provides the Requirement of course. The final Specification was about 150 pages – during its creation we had to learn the dictionary difference between "will", "would", "shall", and "should"! The Commercial Department put a set of words around our thick specification to make it contractual. Engineering predicted the man hours required, from which Commercial predicted the financial needs.

It was a couple of years after the December 1968 AGARD paper that I discovered the real problems of a small 3 shaft engine in a swing wing military aircraft. The three shaft is good for big civil airliners and for their future, but not for small military because of higher than civil power offtake relatively and for very high altitude reasons, so Typhoon's engine had a two shaft engine concept right from the start – its turbines better matched that philosophy too. The small civil BR700 engine series started in Rolls Royce Bristol with BMW sponsorship, is also two shaft.

I was supplying data for the three nation's MoD's and also for British Aerospace, and some internally, so it was late when I investigated the detailed surge margin behaviour of an engine transient, both acceleration and deceleration, which had not been done by anybody in Turbo Union before. I was able to account for mechanical inertia by consulting the Stress Office on their predicted values for each of the three spools, and I put these into the software routines, such that given an acceleration or deceleration the equivalent power offtake from the turbines could be calculated, and added to or subtracted from the compressor powers demanded. Inertia resists attempts to change rotational speed so power is required to change that speed. This power comes from the turbine and results in less power being available to drive the compressor. The result is that acceleration becomes slower and deceleration becomes faster. I decided not to include gas inertia in the compressors (I thought it would be small compared with each spool's mechanical inertia), nor thermal inertia (probably small too, but additionally I had no idea at the time what the metal temperatures would be nor what temperature air the components would be bathed in), and they were too difficult to do in our KDF9 computer anyway. The simulation showed that the engine was prone to sub-idle surge on deceleration, from which it was unable to recover and had to be restarted. I also performed the deceleration simulation with higher and lower IP Spool mechanical inertias, but it made no difference to the idle undershoot loop unfortunately. MTU Munchen confirmed these figures in their own report produced after my work - my good Colleagues at MTU were Klaus Bauerfeind, Jochen Kurzak, Winfred Lauer, and Jon Bjorvik.

The chart below shows the sub-idle loop on the IPCompressor chic. This was a surprise when first discovered, because the power in the HP Turbine was very high (each HPTurbine blade produced over 100 kW at maximum) and so the wing sweep amount seemed small in comparison, but no, it turned out that the IPCompressor surge margin reduced considerably under the influence of such power offtake. The surge margin is that pressure ratio between operating conditions and the surge line, and at altitude the idle speed approached the Max Dry engine rotational speed by design – for wing sweep power offtake was fixed at a high value, but the aerodynamic strength available inside the engine reduced as altitude pressure reduced. This is a significant effect in three shaft engines, but the big civil engines do not have to provide such relatively big power offtake as the RB199 in MRCA had to, nor fly at such high altitude. Clearly the effect of this discovery was to emphasise the need for a good IPCompressor surge line at the expense of high efficiency, and the engine now works quite happily once the ground run wing sweep test conditions are met. However, if the BS143 had been used instead of the RB199, the engine would have avoided the RB199 damaging effects of deceleration and wing sweep, resulting in a simpler and shorter ground run procedure for the BS143. Obviously, in comparison, the EAP and Typhoon prototypes had no significant power offtake requirements and the engine worked happily all the way up to high altitudes, at low airspeeds and in decelerations too. EAP first flew with RB199 Mk104D (Mk 104 as for Tornado F3 aircraft, but without thrust reverse) on 8 August 1986. EAP would probably not have survived design and manufacture without the help of the Italian and Deutsch industry – to their credit.

Thus at all conditions with high power offtake, the IPCompressor running line lifted towards the surge line, above which flow breaks down and the engine drops speed and thrust very rapidly. The BOV (Blow Off Valve) opens at a set HPCompressor speed, and causes more air to flow through the IPCompressor as shown, thus to increase the surge margin. The idle undershoot is typical of 3 shaft engines, but far worse on small military, and is caused by the low slow gain (rate of change of fuel) at Idle relative to the high rapid gain of the fuel system at Max when at 800 knots IAS – this is not so much a problem on big civil engines which only travel at half that aircraft speed. The amount of RB199 fuel flow at Idle can be increased during my defined Ground Run Wing Sweep Checks (that I wrote into the Prototype and RAF Manuals) by adjustment of the VMO (Variable Metering Orifice) on the Main Fuel Control Unit (MFCU), to reduce the amount of idle undershoot during the on ground test wing sweep. This idle undershoot is significant in a cross drive closed situation which occurs when one engine is shutdown, and is less of a problem in flight anyway because the engine swallows more air and idle speed rises with altiude. This idle undershoot occurs at the end of a deceleration, as shown in the diagram.

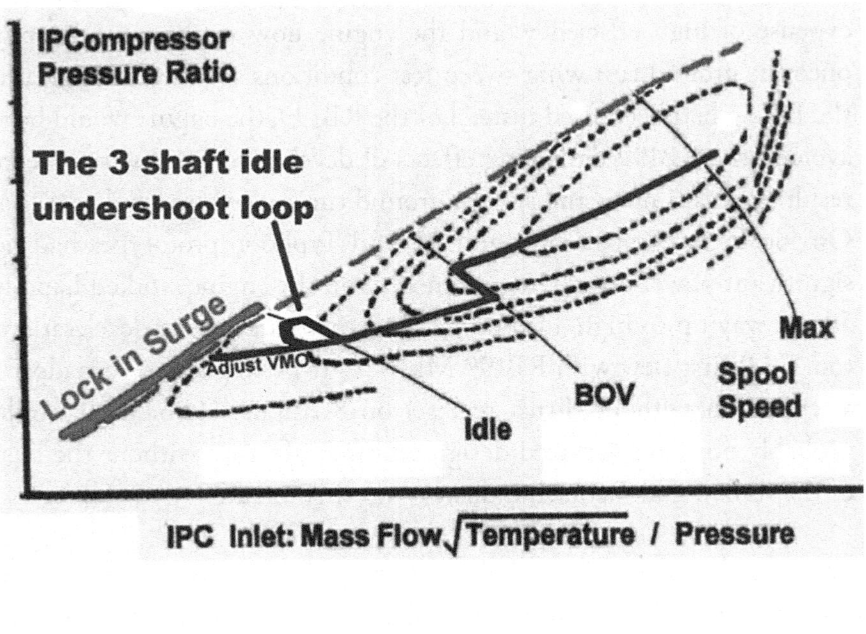

The logic of this deceleration effect is that the HPCompressor running line is lower than the acceleration running line, and the engine will go slightly sub-idle before the fuel system re-accelerates it – at idle this effect reduces the HPCompressor airflow while increasing the pressure ratio because the HP spool is running at idle rpm. The IPCompressor exit mass flow non-dimensional (mass flow times square root of temperature divided by pressure) is lower at surge than it is at normal running, and so the IPCompressor running line lifts towards the surge line at slightly sub-idle conditions.

I am sorry that this description is complicated, but I have found it difficult to write in any other way and my understanding of the phenomenon only came about after 5 years of learning as an Apprentice and the subject is complicated.

I subsequently witnessed this sub-idle surge on an RAF Tornado at the Kemble, Cotswold Airshow in about 1997, when the pilot had to cancel reheat rapidly immediately before take off – I wondered whether the RAF had ignored my Ground Run Setting Up Schedule, which should have stopped this happening! I was also on a Lightning Storm Weather Course in Culham, Oxfordshire when a Tornado in loud fast banging surge went overhead and crashed on the night of 14th September 1989. It turned out this was due to a heavy bird strike and the engines stopped giving thrust. I never heard any more about those incidents, but they were years apart and I wondered if an Expert such as myself should have been personally contacted rather than rely on Rolls Royce plc Departments, for the Original Team would have been disbanded at the test certification of the design.

The software simulation left me prepared for my support of MRCA prototypes P02, P03 and P06 at Warton in October 1974 to December 1975, at which time I left the RB199 Performance Simulation Department behind, after my concept job had been completed, to be the Behaviour Expert in Customer Support.

As a result of

(1) the low thrust of the engine at the time of first flight in 1974 (max dry then was only 93% NH),

(2) the very low thrust of the first engine run in 1971 before any aerodynamic testing had been done (such was the pressure from MoDUK) and the length of time it took to increase the thrust and reduce the fuel consumption,

(3) the sub-idle IP lock in surge problems which had not been predicted until my simulation in 1971, and

(4) the oil-air system problems we had in such tightly squeezed internals – the pilots often had to fly on oxygen because of the rich oil smell, and we had internal fires that consumed oil at a higher rate than normal. MTU (TU Munchen) took on the challenge and cured it.

The procedure adopted for new military engines since, was to design and make demonstrator engines before launching with such an established design, into full scale development. I think the introduction of the three shaft was the cause of all this, for the two shaft BS143 would have been far better achieved, since we knew and could extrapolate the concept so well.

Hence the XG40 programme at Rolls Royce Bristol was the early demonstrator for what became the Typhoon engine, and on Typhoon itself they deliberately designed it to have the established Tornado engine (the RB199) for the first two years or so of Typhoon development flying before using the actual EJ200 engine developed for the Typhoon. Military engines are more complex and are designed for a much tougher environment than civil engines and it does take a long time to get them right. I think that big software projects should also have prototype/demonstrator phases before launching into full scale design and development, for the same reasons.

This flight envelope graph shows Mach numbers for increasing altitude and IAS (Indicated Air Speed). It also shows equal wing drag points for 25, 45 and 67 degrees wing sweep, which I enjoyably and recently computed by hand and without any aircraft data from the airframers. It is interesting to note that they all peak at about 50,000ft altitude below Mach 2.2. Of course an aircraft can zoom from that altitude to reach a higher altitude as MRCA aircraft prototype P02 on its 601st flight to 63,000ft as written in the published Cosford Museum records on the internet.

Note that Tornado maximum speed is 800 knot IAS Mach 2.2, while the F18 is only Mach 1.8 and the F35 Mach 1.6. Typhoon is Mach 2.0, 800 knots IAS.

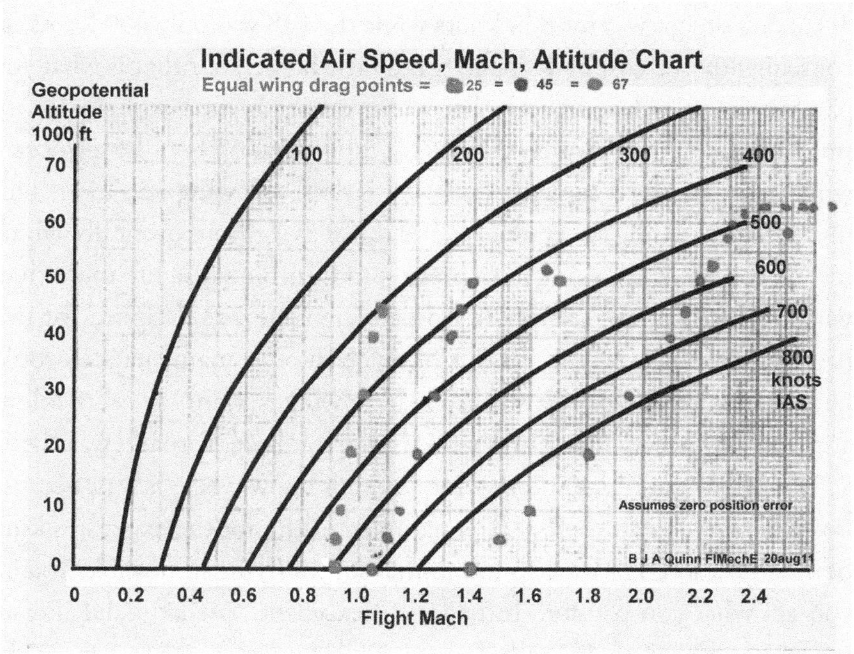

If you compare the Ram Temperature for 800 knots with that for 650 knots you will see that it is 100K hotter - this means you need better materials and cooling for the faster aircraft.

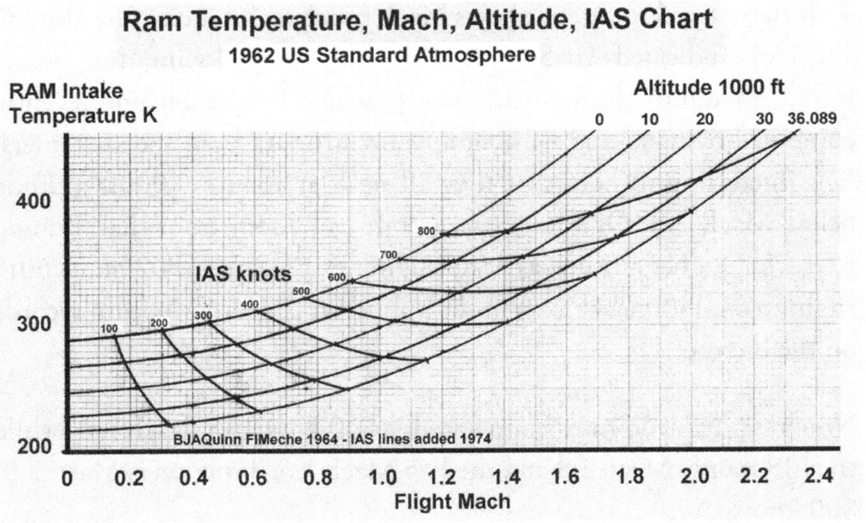

Tornado's design was for 800 knots while the F18 was only 650 knots, so consequently the cost of Tornado was a little higher, but that is what you pay for quality. Always check that what you get is what you pay for. In similar fashion, the F18 was designed for easy operation over the sea/ocean, while Tornado was designed for very low terrain following operation. This involves modulating the engine up and down as it climbs over hills, which means you are consuming Compressor and Turbine disc life more than when flying over the sea/ocean. Tornado's engine was designed for this harsh environment so that its cost of ownership (in maintenance) would have been lower. The F18 also has too big a wing, so cannot terrain follow at speed for any length of time at all. In the Iraq war, Tornado was flying at 50ft altitude at night and the Iraqi did not know where it was, because it was so invisible to their radar - natural stealth, not the expensive kind of the F22 and F35 - the F18 found that sort of flying impossible. Again, you get what you pay for. Tornado was excellent at what it did. Losses in the Iraq war were due to having to climb to release its weapons at the target and so it became vulnerable there. Storm Shadow was developed to strike the target directly at a distance, allowing Tornado to stay down low, all the way there and back, and thus not be seen by the enemy at all. A very successful aircraft/engine/weapon combination.

I worked in the evenings over several days in 1969 to collect the computer output of the extensive simulation for the latest evolution of the design at all altitudes and airspeeds. It took so long to compute those simulations that other users of the computer had to wait until we finished, for ours was the priority to submit information to MoDUK and to the other Nations to persuade them to choose us as the engine for MRCA. In 1970, after more consideration following an MoDUK request to increase Mach 0.9 SL thrust and with an upflowed LPTurbine, I computed a new engine data set and supplied that to the three Nations MoD's and Panavia. We had already worked with MTU on a 3 shaft design by that point, and they were chosen alongside Fiat to finally design the engine in all of its detail with us. The major components like discs and casings had already been launched before contract, for there was enormous pressure from MoDUK to get our engine underway.

MoDUK also said, far too late for it was after the three Nations had selected Rolls Royce's TS1800 (our 1969 RB199-34R submission), that they wanted more thrust at high airspeed Sea Level, which required a significant increase in LP Turbine air flow capacity. The discs and casings had already been launched so the extra airflow had to be squeezed through the same blade annulus as previously. This resulted in a lower efficiency, and therefore higher fuel consumption, compared with all the work that we had done previously to reduce fuel flow as far as possible. All that optimisation work that we and British Aerospace had done to minimise fuel consumption, upset by MoDUK! We were not even able to re-optimise the cycle because of the timescale pressure – there was a panic for some reason in MoDUK around that time, which affected all of us – they even called the aircraft the Panavia 75 anticipating an impossible 5 year programme of achievement. It was eventually 1978 (8 years) before the first production engine was delivered to British Aerospace and 1980 before RAF Cottesmore opened!

During this early "before the RB199 first ran" time in 1971, I visited Saab in Sweden with our Chief Engineer Peter Orchard to talk about the engine for their new small fighter. We flew in the Company HS125 with two others I think (one from Design and the other from Commercial Department), but unfortunately we were turned down later – I think that GE won that competition financially with their F404 engine, which now powers the Saab Grippen, a very useful single engine canard fighter. I think the RAF should buy some of them for they will be cheaper than Typhoon and yet have good performance in a fight. We had many Hurricanes winning the Battle of Britain in 1940 and I think warfare today should think in numbers terms not just the highest (expensive) F35B technology.

I later also visited Belgrade in Yugoslavia to talk to them about their LCA (Light Combat Aircraft) with our RB199 presentation that I took there. I was floored by a question from their aerodynamicist about "how does a bumble bee fly" because I did not know. It was a myth that it broke the laws of physics! I now know that bees fly by rotating their wings in a circular motion, and it creates pockets of low hurricane-like air pressure above the bee's wing which lift it into the air and thus grant it the ability to fly. So, the normal laws of physics do still work for them!

In 1987 I visited the Pentagon to give them a briefing on the RB199 engine so that they might consider it as a replacement for theirs in the A10 Warthog – a very forlorn hope, promoted by the Marketing Department inside Rolls Royce.

One problem that gave the Russians our technology was the 1970 release to the media of the choice of bypass ratio (which is the ratio of bypass duct airflow divided by internal compressor airflow). I had given our Installation Engineers the information about top dry (non-reheated) temperature at the tailpipe and they had computed the Infra Red signature as a result – necessary to know for anti missile purposes. Releasing the bypass ratio that early meant that the Russians could now refine what future missile sensor sensitivity they needed to bring Tornado down, through their own predictable computations. The Performance Department should have been consulted about what information could be released that would not help the enemy. That bypass ratio should never have been released to the media in my opinion, but it was too late, it was already out there. The other parameter that automatically comes out from bypass ratio is the fan pressure ratio given a probable HPTurbine entry temperature – this gives the Russians all they need to know about the engine compressor pressure ratios too, so any classification appropriate to those is worthless! Given fan pressure ratio and an equally obvious approximate overall pressure ratio, the HPCompressor pressure ratio comes out from what the HPTurbine can deliver and then automatically it follows what the IPCompressor pressure ratio is – hence all you need to know is bypass ratio, and the figures for pressure ratios do not have to be that accurate to determine fuel consumption for example. I have told Rolls Royce and MoD of the future need to ask Performance Department to decide what can be released about a military engine, that would not help an enemy, and not to release bypass ratio.

I saw Sir Frank Whittle walking in Bristol Siddeley's Whittle House foyer on his own, and grinning at me for no reason other than his own cheerfulness, and mine too. Sir Stanley Hooker (SGH as he was known in Bristol) was very approachable, listening to several young computers like me (at the age of 29 back then on the MRCA engine) in his Rolls Royce Bristol office at the time of the Rolls Royce bankruptcy in 1971, when he was asked by the Government to review all of Rolls Royce's engine programmes. I went to Air Commodore Sir Frank Whittle OM KBE CB Thanksgiving Service on 15th November 1996, at which the RAF reserved me a prime seat in Westminster Abbey, and saw the flypast of four Tornados from 111 Squadron and two Meteors over the Abbey, from the North Green.

Sir Frank Whittle invented the jet engine in 1930 when he was a pilot on a training course. Despite limited funding, a prototype with a large diameter centrifugal compressor was created, which first ran in 1937. By 1941 Rolls Royce had been persuaded to join in with the development by Sir Stanley Hooker, who observed that the early jet was already producing 800 lb thrust, which was as much as the Spitfire engine propeller produced, and that it had considerable potential therefore. Later in 1943, the engine was producing 4000lb thrust.

The Deutsch meanwhile had seen Sir Frank Whittles patent, and launched their own engine through design and make, but they adopted the smaller diameter axial flow compressor. The Deutsch Engineer's idea was supported by Hitler's Government and was designed as an overall big aircraft project concept from the start, while the UK stumbled over the idea, even though Sir Frank Whittle's jet engine had already shown exceptional performance compared with the best piston engines of the time. So the UK Government delayed our potential fighter by choosing to design and make a rather simple demonstrator aircraft (the Gloster E28/39 which flew for the first time on 15th May 1941) instead of an all out fighter concept at that time. I think this must have been because they had no Senior Engineers in their procurement department, and thus UK Government obviously did not understand how to plan for, nor appreciate, the best technology in war.

The Deutsch Me262 aircraft engine was troubled with many axial compressor blade failures, but even so became quite the scourge of allied aircraft, shooting down over 540 aircraft. The Meteor had its first flight in March 1943 and saw minimal action in the war, though it did have some success against the Deutsch V1 missile bomber, for it was fast enough.

The Gloster Meteor gained world speed records after the war, but that I think was because the faster aerodynamic swept wing Me262 was denied its proper place by the UK Government. The swept wing was copied by the allies on other aircraft even so, for it was such a good concept, and is standard on all military and civil aircraft today.

I think the very slow Lancaster should have been superseded by a jet powered bomber much earlier than the Canberra was, and that and the Meteor could have been launched at the same time as the Gloster E28/39. Perhaps the UK Government never understood the need for speed, yet that must have been evident as a result of all the bomber losses against the much faster fighters right from the start of the war. 55,000 men and women lost their lives in the RAF's Bomber Command during the second world war!

In my view the UK Government lost the technology war, for it seems it never thought about our long term survival, just plodding on hopefully. Having found a Lancaster they stuck with that old concept, instead of thinking – it even took them a long time to adopt the very capable and fast Mosquito designed by de Havilland. It was the huge contribution of USA and Commonwealth soldier numbers that saved us. What a good job we had radar for it helped save us in the 1940 Battle of Britain, where our good fighters were faster than the Deutsch bombers – the go faster lesson was not learnt by the UK! It was a good job the war ended quickly, for the Me262 could well have killed even more of our aircraft and crew.

Today we are on top of the best technology in Tornado and Typhoon, so the technology lesson has now been learnt – I hope this applies to the Army and Navy too.

RB199 FLYING

I was one of the three people (Dave Wellings – dry, Dave Tiley – reheat and Me – surges and performance overall) who regularly worked with the Vulcan XA903 Flying Test Bed team led by Harry Pollitt (later Graham Andrews and then Ron Cowpe) the RR Chief Test Pilot, and Alan Baxter (lead Flight Test Engineer) to plan and analyse the Vulcan's work. The Vulcan had the correct intake and SPS (Secondary Power System) gearbox from MRCA, but was unable to take much power offtake, which proved critical to Tornado operations, and which therefore led to the early demise of the Vulcan's usefulness. Among many tests carried out on reheat boundaries and so on, it was shown that the cold relight boundary met the Specification Requirements, but more interestingly that an immediate relight worked all the way up to over 40,000 feet altitude subsonically! Only 25,000 feet was expected! I also learnt that a hot day at ground level meant a cold day at altitude in the tropics! The Vulcan also gave us the confidence of flying MRCA for the first time, especially when Paul Millett, the British Aerospace Chief Test Pilot, came down to fly the RB199 in the Vulcan with Harry one day.

During the MRCA ground run procedures before first flight, I was told by British Aerospace Engineers that the wing sweep power offtake on each engine was larger than I expected. The engine had surged on slam decel from max reheat to idle with the wing sweeping and the cross-drive closed so that the single engine was delivering the whole wing sweep power offtake on its own, and it thus was to be avoided during flight. I therefore later used the software simulation when I returned to Rolls Royce Bristol from BAe Warton, to look at the maximum possible altitude that a potential production engine would be required to deliver with wing sweep with cross drive open, to discover that it was over 50,000 ft. at 200 knots IAS if our prediction for compressor aerodynamics was correct. This information was received with no immediate comment when I presented it to our British Aerospace colleagues, but it was acceptable. This flight condition was also the limit for reheat modulation, but these figures should never have been released in an English language Paper presented in Japan by Turbo Union in 1983, for the Tornado was only just then being introduced into European Service. Did MoDUK agree the unclassified release? –

somebody needs sorting out! It is no longer of concern since Typhoon now does the job at that altitude.

British Aerospace had defined the full range of checks they wanted to examine before P02's first flight, and from that I created the engine specific ground run procedures (for the RAF manuals later) with British Aerospace aircraft specialists helping to define their aircraft interactions, to certify engines for flight in any future installation need, and for various engine and fuel system repair and installation needs. It was during my early visits to Lancashire's Warton that I thoroughly enjoyed the taste of mushy peas, which has stayed with me ever since!

MRCA XX946 was P02 whose first flight on 30th October 1974 was effectively my second experienced flight as per the extensive recordings I had been given for MRCA D9591 (P01) first flight on 14th August 1974 at the MBB airfield at Manching near Munchen. I had not been at Manching for that flight, but I had been given a magnetic tape (processed by our Computer Department via Brian Banes Rolls Royce Bristol software, and supplied by MBB Flight Test in Manching) of all relevant engine and aircraft instrumentation. I could find no reason for the P01 pre-flight surge from P01's magnetic tape, but all of the flight recordings were well controlled, and no surge occurred during either the actual take-off roll or the flight afterwards.

On P02's first flight, they had a surge preflight and I was called into the Telemetry Room by Harvey Francis, the TU Chief at Warton – he had a team that I knew well, Brian Gregg, Roger Weaver, John Maskell, Jim Sweet, Les Summerfield, Chris Brain, Richard Jaggard, Bob Studley, and Jeff Stores, for example, and all good friends. My instructions to Lieutenant-Commander Paul Millett OBE DSC Queen's Commendation for Valuable Service in the Air, were to explore the surge margin of the HPCompressor by accelerating hard while I knew the surge margin was at its smallest at the top NH, and the rapid deceleration check was to explore the IPCompressor surge margin.

The fan surge margin would have been explored in both cases there being no difference in the expected running line on either acceleration or deceleration, while the preflight surge had been at Max Dry. The engine did not surge during those checks, so I was happy that the engine would be OK - after all, this was my second flight, just as it was Paul's with a preflight surge – and so it turned out to be, for it behaved well during the whole of P02's first flight, just as P01's engines had done on P01's first flight.

After my instructions from the telemetry room to check out the P02 engine were performed satisfactorily, I confidently said that we should go for flight - "OK, go for it" I said, but this was poorly reported by Dave Williams who was acting as "boffin" on the communication microphone to the aircrew. Dave simply said "over to you Paul" when he should have said "we are happy, over to you Paul". Paul Millett and Pietro Trevisan from Italia (in the back) had gone offline briefly following Dave's handover and then said they would go for flight too.

Immediately after the flight I told Wing Commander Roly Beamont CBE DSO DFC DL FRAeS (British Aerospace Director of Flight Operations) that it must have been a warm up surge problem, and I then created the warm-up procedure (I tested my idea the next available P02 day) that was used on all the prototype MRCA-Tornado before flight – it was to accelerate to Max Dry for 5 seconds, throttle back to 90% NH for 30 seconds, then to max for take-off. This worked well for the next few years, preventing pre flight surges. I thought it was the back end of the HP compressor that had triggered the surge because the surge margin was at its minimum there and the blades were tiny – rapid growth of the HPCompressor casing and slow growth in the compressor discs had led to a temporary increase in blade tip clearances, which closed down again after warm-up. Larger tip clearances mean lower surge line. Xray evidence a year later confirmed this. I thought that fins on the outside of the HPCompressor casing would be required to keep the casing cooler by bathing them in the cold bypass duct air.

During that taxi out to first flight, Paul Millett complained that the taxi speed was too high at over 70 knots. I discovered from my software simulation later, that thrust at idle would be halved if the nozzle was opened up to the max reheat area, so that idea was adopted. It took some time because the MECU (the engine electronics) had to be modified to accept a cockpit switch input and to then demand the wide open nozzle. The switch had to be incorporated into the aircraft cockpit too. Both of these took some time to do, but when done, the taxi speed came back to a much lower value as predicted – and all was fitted to the Production Tornado.

The other problem noted by Paul Millett, in his first flight debriefing which I attended, was that on one occasion one engine had not quite met Max Dry when selected. I looked at the telemetry recordings afterwards with my friend Peter Dootson (who was with me from Control Systems Department for first flight), and the MECU milliamps was near zero indicating that the engine was still trying to accelerate, but could not. It seemed the fuel system was limiting, so Peter asked Dave Wellings in Bristol Controls what to adjust. We did this on the first available P02 day, and thus corrected the situation by using ACU, DCU and VMO adjusters, which I then asked Rolls Royce Bristol's Technical Publications Department (Gordon Richards and John Byles were particularly good) to put in the engine manuals.

On P02 flight 7, an HPTurbine blade was discovered by borescope inspection to have failed at half height. Paul Millett did not know that it had failed, and he asked me what indication there was from the recordings, so I plotted the data out by hand for that flight and eventually discovered that the only indication was less than a quarter percent HP spool speed reduction, there being no significant vibration level change. Lieutenant-Commander Dave Eagles AFC (ISBN 978-

0-7509-6841-6 "Testing Tornado") had already commented upon his first trip in MRCA that the engines were very smooth, and that he could feel no vibration at all, and that was Paul's comment after that blade failure flight too. Both Paul and Dave, ex Navy pilots, thanked me for my Engine Behaviour help to them during the early years of MRCA prototype test flying, and my reputation with BAe Engineers was good too.

As a result of a bird strike, and thus a failure to light reheat on P02 Flight 18 in April 1975 which was not recognised for a short time during that flight, I sat in P02's cockpit in the hangar afterwards while our ground test nozzle plate actuator was fitted, and I moved the throttle myself into the reheat range, to see the area open up on the cockpit gauge, but not of course to see the engine spool speeds for the engine was shutdown at that point. I was not present in the telemetry room during flight 18, and only seeing those spool speeds would have shown that reheat was not lit, but I saw the recordings afterwards. Dave Eagles in the rear cockpit had commented at that flight point that the HP spool speed was low and the LP spool speed very high. This was the indication that reheat was not lit, and I created the spool speed specific telemetry software analysis routine for BAe Warton in April 1975, to show reheat "lit" or "not lit" advice to the telemetry room crew, and thus to MRCA-Tornado Test Pilots in flight. I also worked out that a wide open nozzle at maximum spool speed with reheat not lit, would halve the thrust available – half max dry thrust is not enough to get home in the single engine case, so later the Turbine Pressure Ratio Transducer was introduced to detect a reheat not lit situation, actuated by the MECU electronics to automatically close the nozzle. This took some time to do as the MECU had to be significantly modified to make that response work. All of these modifications were in the Production Aircraft of course.

During my time in the Performance Department I had also defined the idle schedule which was subsequently implemented in the MECU

(Main Engine Control Unit, now called the CUE – Control Unit Electronic). Because of the considerable problems of the three shaft we had decided to increase the idle speed as altitude increased. At high altitude the idle speed was thus equal to the max dry speed, while to avoid intake buzz the idle speed had to be max dry speed above Mach 1.3. Intake buzz is a violent resonance in the aircraft intake which can occur at high Mach for all supersonic aircraft including Concorde – thus the reason for movable intake ramps above Mach 1.3 on these types of aircraft. Interestingly on one flight at medium altitude, a development Tornado could not slow down from Mach 1.2 without pulling G to increase drag. So, I had to reduce the idle speed a touch for the clean aircraft – Tornado was super-cruising! The idle speed was a function against altitude ambient pressure and Mach number measured by the TTU (Triplex Transducer Unit) and driven by the MECU of course.

I was also responsible for defining how to measure engine Thrust In Flight, so that Panavia could check the drag of the aircraft in its various configurations, and I went to MBB Manching to talk to Egon Hinz and Hans Fichter about that correlation, which could be used for future developments of the RB199 too. Egon took me to see Don Giovanni in the Munchen State Opera Theatre, and he did some translating for me, for it was in Italian! A lovely theatre and a great performance.

My favourite P02 on the ground picture is at

http://www.tornado-data.com/history/prototypes/p2.htm and a good one exists on the front cover of "Testing Tornado" ISBN 980-0-7509-6841-6 by my friend Dave Eagles AFC where P02 is at Mach 1.6, 620knots at just over 30,000 ft with a single seat Lightning taking the picture, quite bravely closely following P02 at such a speed with one hand on the camera and another on his own controls! The Lightning had a maximum speed of 650 knots so it was virtually flat out to stay with P02, capable of 800 knots at such a height!

SEA DART

This Mach 3+ missile was Bristol Siddeley Odin ramjet powered and its early flight performance had been monitored by Hawker Siddeley, who found that the range was a little shorter than expected, so I was asked to look at all of the engine test results and work out how to improve the fuel consumption calculation, by investigating our combustor efficiency correlation. I used a plot of efficiency against Longwell Loading (a standard industry combustor characteristic) for each engine and then used two dimensional polynomials to determine the characteristics. I started with 3rd order polynomials but found that was not good enough, so I increased the order several times until I found that 7th order polynomials worked just fine across the whole efficiency and Longwell Loading range for several engines. 7th order means that x starts to the power of 7, then 6, 5, 4, 3, 2 in the long equation. These equations I supplied to Hawker Siddeley's Performance Engineers. Each Odin engine (whose High Altitude Test Plant measurements were already known to Hawker Siddeley) was calibrated against those polynomials by adjusting the efficiency in a simple delta from the 7th order 2D polynomial correlation, and so I had done the job. I heard nothing back from Hawker Siddeley so that worked well for them.

CHAPTER 3

MRCA-TORNADO CUSTOMER SUPPORT

In mid 1975 I applied to and then joined the Customer Support Department in Bristol and became Assistant Service Manager RB199 after a couple of years. I did not go there to be a Manager at all, but I did want to be near the aircraft and its needs, having spent sometime with the engine test, assembly and repair crew at Turbo Union's works at British Aerospace Warton facility in Lancashire, with my background in Engine Behaviour. I also wanted to prepare for and understand the Operations needed by the RAF, GAF, GNY, and IAF as Tornado entered Service. The RAF would be the lead operator of course, so were likely to find the first problems which I and my team would have to deal with. I wanted more hardware exposure, but I also thought the job I had been doing inside Performance Department was running out, since only engine test development was needed now – not for me.

I worked with one of the RB199 team called Dick Truebody – he was fairly ancient by then having served as a Bristol Rep (in Cuba in about 1960) for the Bristol Centaurus engine in the Hawker Sea Fury aircraft. The Bristol Centaurus was the most powerful piston engine in the world at 2500 HP, far more than the RR Merlin's 1300 HP. He told me that one day he was running up the engine having tied down the Sea Fury with ropes, demonstrating the ground run procedure to Fidel Castro. As Dick brought the engine up to full power, suddenly one of the two ropes broke. Fidel Castro was standing directly in front of the aircraft, and Dick could do nothing other than throttle the engine back, but that was after the aircraft had swung around nearly decapitating Castro on the way! So, Dick nearly became very famous suddenly – no wonder he was so cheerful when I met him, for he survived the experience!

I learnt about non-specification needs like the way engines were handled by Pilots when aircraft were taking off in formation, and aircraft operations interaction with the engine. For example, I identified what needed to be done to calibrate the engine on high altitude static ground run checks, or what to do if you hear reheat buzz where I wondered how to detect it, so I created/identified the musical piano key 19 (D♯/E♭) which closely sounded like buzz, and which could then be used by the ground run engineers at any Tornado operational station to calibrate their ears before the ground run. I defined and wrote the ground run adjustment procedure that may be necessary if it buzzed, from a knowledge of the test bed discoveries by Test Engineers. Clearly ground run tests and adjustments provided the important clearance to fly, and should not be altered merely to reduce run time. No RB199 Mk104 engines failed due to reheat buzz which can trigger failure of the LPTurbine Shaft (ref ISBN 978-0-7524-5936-3 "Tornado ADV"), though because of their longer tailpipe they were more susceptible to buzz than the Mk103, and were the reason for introducing the piano note frequency check.

I also introduced a small hand held Hewlett Packard HP-65 programmable calculator that had just come onto the market in 1975 – they were £350 each. This small computer allowed me to programme it and record the programme onto a magnetic card. So, I asked for funding to provide several of these for the Service Engineers supporting MRCA around the UK. I then sent my write-protected programmed magnetic cards to all of them so that they could compute the settings required for the open loop reheat while they sat in the cockpit. The output gave the settings differences required from those already set in the MECU during their reheat ground run. Reheat was initially setup according to the test bed results, but always needed adjustment in the aircraft at several points up the reheat function from min to max reheat, and according to a corridor I established against the LP Spool Speed drop from Max Dry Speed. The sensitivity of reheat fuelling up that range varied and it was easier for the Service Engineers to use with my programmed magnetic card, than for them to work it out for themselves from the Engine Manuals. I so advised the RAF later too, and they procured their own computers for their own Ground Run Engineers.

Open loop reheat means that the fuel flow is scheduled according to the flight condition (altitude pressure and temperature and Mach Number), whereas closed loop reheat is where the running line in the engine is set and the reheat fuel flow is adjusted to achieve that running line. The difference that results is that the open loop (if you can set the schedule accurately enough) will give both the highest thrust and the fastest modulation to max reheat, compared with the closed loop which can cause over-fuelling and thus quenching of the flame, causing loss of thrust and high fuel consumption – the Phantom Spey was closed loop and suffered from this. On the MRCA-Tornado engine we used open loop, based upon these arguments originally put forward by Klauss Bauerfeind at MTU. We then used the TPRT (Turbine Pressure Ratio Transducer) to monitor the engine running line and cancel reheat if the running line went too low, indicating flame-out. It works well today.

I supported our AOG team (Aircraft On Ground team in Bristol who existed to deliver goods to Customers Aircraft waiting on the Ground) by taking equipment and components in my car to Warton or RAF Cottesmore when I went for meetings there – my car was appropriately insured of course.

I attended the 1976 Farnborough Airshow on the first Monday as a Service Engineer. I was on the opposite side of the runway from the crowds, and I was far enough along the runway by the trees, to be under the flight path of Paul Millett's solo display in MRCA P06. Paul took off and pulled G in a hard right hand turn which took him right over my head at about one hundred 100 ft altitude when the left engine surged with a big bang. I thought "that's my mate up there" and watched him continue his display for a few minutes as if nothing had happened. When he landed, I walked over to him as he got out of the aircraft, (he was quite calm!) with the other Service Engineers to talk to him, but we agreed it would be better in the BAe tent. The result was that I volunteered to go to Warton with the aircraft magnetic tape recordings to perform some analysis on Warton's computers. BAe shuttled me into an aircraft to Warton, where I worked with Bob Hartley (BAe Flight Test) to review the recordings. I found that the engine was good before and throughout the flight, with all recordings as expected apart from the surge, while Bob found that Paul had pulled very high AOA (Angle of Attack, or incidence) at the point of surge, and not afterwards. We decided to go for a ground run next morning, just to be certain all was OK, and so it turned out. The rest of the week continued without engine change quite satisfactorily. I had gone to Warton without an overnight bag, but I met Barry Miles from Bristol Control Systems in the hotel and he gave me a length of toothpaste for the next morning!

After the Farnborough week, it was decided by BAe and Darryl Williams, to swap the left and right engines over, and fly high AOA again. The left engine surged as before and the right was OK, so it was a left intake/engine interaction problem. I was much involved in this, as were our Installation Aerodynamicists, my friends, Darryl Williams and Bob Hercock. ISBN 0-9530345-0-X "The Birth of Tornado" says that they had to put a small fillet on the left intake to cure the problem.

Single Point Curve Integrity

During my MRCA concept days, I constructed many parametric studies to help me understand the engine functional behaviour. The result over 4 years or so was that I knew what affected what, in detail, and thus the characteristic functional shapes. Appropriate simulation studies are useful in any activity.

When it came to the real engine in the real aircraft, I was thus able to rapidly understand what was happening. I extrapolated for that environment too, and was always able to spot a problem in any specific test as a result – if you know what to expect, and it does not turn out that way …….. Given a measured test point, I was then able to fairly precisely draw the function. This technique led to the description "Jim's single point curves", and later test point results nearly always confirmed that what I had drawn was accurate. Thus my reputation improved. Indeed I was able to construct such functional curves, even to the point of describing how the engine might behave in an aircraft spin – BAe Flight Test's Bob Hartley (the engine specialist), said that I had identified the character of a spin precisely – he had model data that I did not have, to so congratulate me. I based my thinking upon the falling leaf, and flight tests of other aircraft, but I did not know of their engine behaviour. It is essential to keep your eyes open to search for as much data as possible and think. I learnt that a pilot will find a spin rate of 2.5 radians/second (150 degrees/second) very difficult to control, so P02 was equipped with a rocket motor when spinning trials started to help recovery.

My accuracy of advice was important to me – it gave me credibility within the team – the whole MRCA-Tornado team, not just within RR. My RR boss, John Allan, had confidence in me based upon his experience of working with me, and always knew that what I said was genuine – he said "don't change Jim". He meant "do not start playing with words, keep them clearly honest and simple, which I am luckily", and of course that wisdom supports me greatly still.

I did not change, because what I said affected the flight safety of the aircraft, let alone our reputation as a Team – a Team I wanted to belong to, and had pride in. I enjoyed the responsibility, and giving advice to help others. I liked coordinating the efforts of many people too, and learnt a lot thereby.

I wanted the aircraft to be successful, to be a deterrent to any enemy of any of our Tornado Operators – and not because Tornado initially carried nuclear weapons, but because it was an accurate deliverer of non nuclear weapons.

I am a Design Engineer, and take pride in my integrity and honesty. And my career thus helped me become independent.

Small Test Bed at Warton

John Allan was Service Manager and my good and immediate boss. As a result of a request from Mike Henney, the British Aerospace Flight Test Chief at Warton, who had asked John not to use the life up on his aircraft components (!) just for engine testing purposes after modular rebuild on site, asked me to commission a small test bed to be used at Warton for engine testing. I went to Cardiff to see the people who were designing the RAF's RB199 test bed, and they quickly got themselves into gear and produced a simple test facility for use at Warton in their Dutch Barn. This successfully saved the aircraft from being used as a test bed after engine module change on site, saving interference with the aircraft flying programme, and saving life consumption on aircraft components.

Organisation

Our organisation was so integrated with the RAF's that I was able to produce the chart below showing our relationships, and I later received a congratulations letter from the RAF addressed to Director Bristol, John Wragg. I visited RAF operating bases several times each – Cottesmore and Honington particularly and once each to Lossiemouth, Laarbruch, and Bruggen.

Part of our job in Customer Support was to produce the International STO (Special Technical Order which was followed in our UK case by the RAF STI and SI) that our Customers needed to maintain their engines. I had invented the TUPSI (Turbo Union Product Support Instruction) procedure that all three nations would use on the aircraft and engine build lines, and we had used that procedure, signed by the Bristol REO (Resident Engineering Officer employed by MoDUK) in the UK's case, to instruct work at the engine line in Warton, where that work was necessary before dispatch to the RAF in a Tornado aircraft. The STO procedure was agreed at NAMMA level with us Contractors (Panavia, Turbo Union, Mauser), and I started the whole system off in Turbo Union with the first STO which was to sort a problem with the IPBOV (IPCompressor Blow Off Valve) in July 1980. I personally produced most of the earliest STO's, to a total of 60% of all that were produced in the first five years of RAF Operation with Tornado, ranging from surge prevention, to electronic resistor checks in the Lucas supplied Engine Electronic Unit, to clip security on fuel pipes, to cable loom fretting, to eliminate reheat buzz occasions which might possibly caused LP shaft spline damage. What came through all of this was the excellent honesty and integrity of all those who worked on the Project.

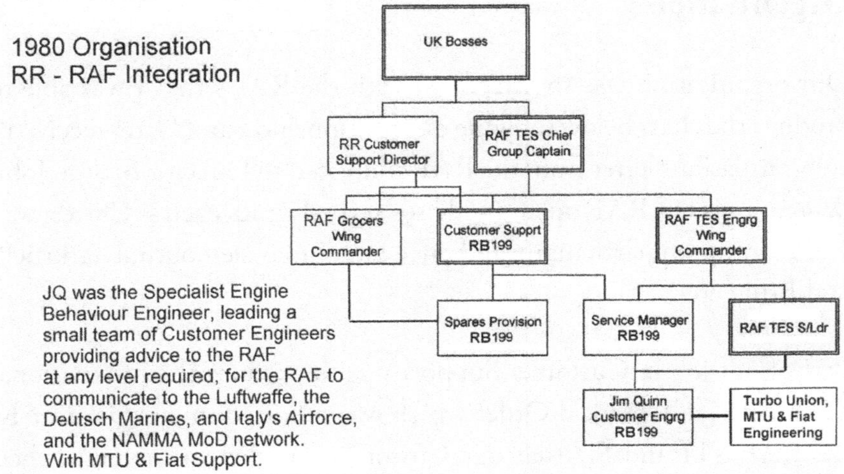

REPAIR NETWORK

John Allan also asked me to identify the reject repair process with RRAnsty Repair Department, and thus to launch the Tornado Engine defect and repair communications network within RR Bristol and RR Ansty, coordinated with our partners in Europe. RRAnsty were delegated by the Rolls Royce Repair Director to repair MTU's design components that the RAF rejected, and they were very keen to support the RAF well. I went to RR Ansty to discuss how we might work together, and found that they wanted us to tell them why a component had been rejected, because that would help them to plan and carry out their repair work at lowest cost. They were obviously already being kept informed by MTU Munchen about how to strip and rebuild the components and what to order as necessary based upon dimensional limits. This conversation resulted in my team at RRBristol Customer Support talking regularly to RRAnsty on paper about the RAF rejects that occurred. This resulted in a close collaboration and the first benefit was that RRAnsty found that the abradable coating inside the MTU Mk101 HP Compressor casing was being pulled off by plucking dimples, so we thus worked out the reason for the early surging, in a correlation report.

This report went to the FG4 Chief Engineers Design Audit Functional Group and thus MTU were actioned. Dr Jost Schmidt from MTU Munchen, was the good Turbo Union Chief Engineer who over saw the Chief Engineers at Rolls Royce Bristol and Fiat Torino. MTU identified that blade passing frequency was the cause and changed the resonant frequency of the abradable by changing its designed thickness – a modification that was rapidly introduced into the fleet by both RRAnsty on RAF components and MTU on the new Mk103 components. I learnt about the economics of the factory while in Customer Support as a result of being exposed to spares predictions by colleagues. If, for example, the orders on one engine type went down, then, because the factory cost is the same, the prices of orders for another, stable, engine type went up, unless the factory swallowed the loss of profit to avoid losing orders for it, or people were made redundant to reduce factory cost.

There are four levels of maintenance in the RAF - 1st level is at aircraft dispersal ready to go for flight, and you want to make sure it can. 2nd level is in the hangar at an RAF operating base where aircraft or engines were part stripped and rebuilt with components and modules available as new or reconditioned spares. 3rd level was RAF St Athan where modules and components were needing possible machining to put them right, or at least had the availability of new spares from the factories in Turbo Union. 4th level was RR which went to the same (or deeper) depth as RAF St Athan. Terry Shore from RR Bristol Development Engineering provided us in Customer Support at John Clarke's meeting that I attended too, with reliability estimates based upon development testing. These estimates were passed on to both our Spares Ordering Department under Derek Wilding, and also to Wing Commander Dave Saunders who estimated the number of engine maintenance stands required to strip and rebuild engines at 2nd level. There were quite a large number of them and the space needed was considerable, but that was Dave's job, and RAF Cottesmore never had a problem with engine availability for training all the pilots and navigators from across Europe. The training courses obviously had to be on time, for otherwise the UK would have a bad reputation for delaying all of Europe's training. However, it also turned out that the engine was more reliable than such early Development estimates, and it turned out that RAF Cottesmore did not need to have so many engine maintenance stands after all.

WEIBULL

I confirmed that the weibull distribution (a log log computation as pictorial graphs shown below) statistical analysis techniques most reasonably represented the observed RB199 component reliability environment in Service, for two whole engines in a wholly integrated aircraft in a rapidly changing flight condition, as against one engine in an old brick built static ground test bed. The technique was named after Swedish mathematician Waloddi Weibull who described it in 1951. The first chart is what we plotted from the available data, using straight lines to go through the points that the computer people upstairs gave us from the available RAF data – they collected all the module flying hours, which is why they were abler to compute the Weibull probabilities of each module. The chart underneath shows the interpretation of those straight lines in the more familiar non-log-log failure probability chart.

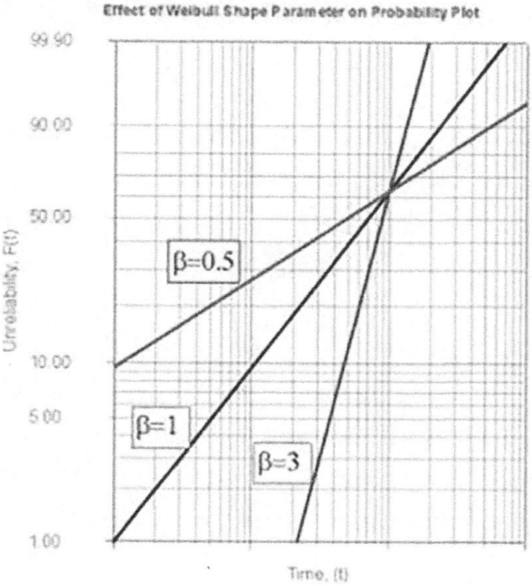

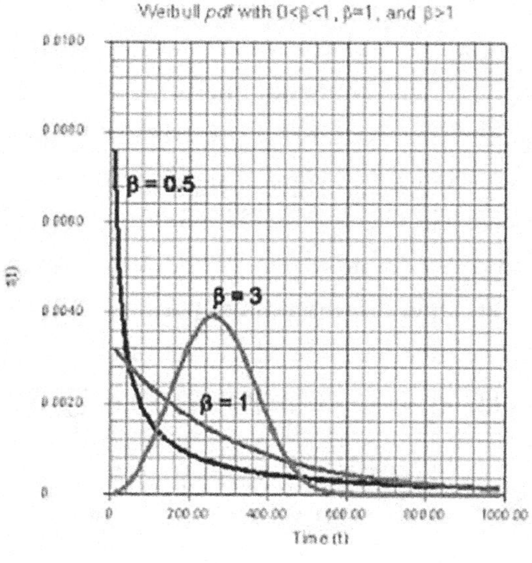

The RAF were lead operators of Tornado and they were therefore the first to suffer Mk101 IPTurbine blade failures, and RFCU (Reheat Fuel Control Unit) failures amongst other less important problems. I asked my Customer Support computer colleagues upstairs to produce Weibull probability's for me on the RFCU which I plotted and thus identified the risk of failure associated with various lives of components in service.

The REO (MoD's Resident Engineering Officer – originally the good Don McIntyre for most of the development programme) was another good man, Roger Tasker, and he ran regular meetings after 1980 (RAF Cottesmore's TTTE establishment date: Trinational Tornado Training Establishment) in Bristol with Rolls Royce Bristol, AQD (MoD's Quality Department representative, Eric McRavey), and an RAF Officer (Squadron Leader Mike Moffett MSc Metallurgy, later Wing Commander, from Whitehall, London TES office – Tornado Engineering Support) representation to review progress on the engine and its In-Service behaviour. My reputation with them was good.

At the time when RFCU failures were becoming more evident, there was discussion at the REO's meeting, about how to control the failure rate and whether the RAF (and the other Operators – the GAF, GNY, IAF) should adopt a life limit, but I presented a Weibull chart for the RFCU which showed a Beta Value of significantly less than one. This meant that the problem was an early life one and that once the RFCU had gone beyond a certain number of hours life, it appeared to be able to go on for a very much longer time without failure. The problem I concluded and argued was therefore manufacture of the very fine tolerance metering valves. Thus I was thanked and it was agreed not to have a life limit on the RFCU at all. My argument allowed the Engineers at Dowty, who had designed and made the RFCU, to concentrate on what was going wrong during manufacture and I expected that they would find the problem and put it right.

I asked my Customer Support computer colleagues upstairs to produce weibull probability's for me on the IPTurbine blades which I plotted and thus identified the risk of failure associated with various lives in service, and my plot showed a high life probability value and it had a Beta Value of about 3 (indicating wear-out – see charts). I communicated my recommendation for that high life limit to the RAF Engineering Office in Whitehall and they arranged to assess the whole fleet and to move engines around quickly, so that one high life engine was alongside a lower life engine. Those engines with flying hours approaching my Weibull predicted limit were removed for replacement of the IP Turbine blades. Thus the failure problem did not recur, and the fleet was not grounded.

The reason for that all of that removal work by the RAF, was because that Mk101 IPTurbine blade failure caused an engine shutdown, and thus a double engine failure might result in the loss of the aircraft if both engines were at high lives, and we needed to ensure the safety of the fleet without stopping Operations. This removal action was successful and no further IPTurbine blade failures occurred. I wrote the Customer Support Red Top Paperwork which highlights to all the RR Directors and to MTU and Fiat when there was a big problem. MTU who designed the IPTurbine blade also now did not have to panic over a grounded fleet, so had time to sort their design out. It should be noted that development flying investigates particular parts of the aircraft's exposure to the environment, and does not fly Customer Operated Missions to determine the likely failure rates of components in Service, because it would be so expensive to cover the whole range and number of possible missions, some of which were undefined during development.

The Customer Operates several different types of mission, and TTTE Cottesmore (the first user of Tornado in Service) merely flew training missions. So Service Exposure reveals problems that cannot be found in development. They have to be managed, without grounding the fleet, by Customer Support thinkers like me, in consensus agreement with the Customer Operator. Customer Support also discusses the problems with the Design Engineers to ensure the future designs adequately meet the new discoveries, and I used to ring them fairly frequently, including to MTU's Adolf Wulf (the TU Chief at MBB Manching, who I visited once with RR's Stan Yeandle) on thrust reverser changes early on P02's flight test programme – Adolf made things happen, and P02's back end modifications were very quickly supported by MTU as a result. Harold Joerg and Lothe Kunze were very helpful in MTU Customer Support too. It was this level of help and friendship that led me to vote REMAIN (with other reasons too – see later) in the EU Referendum election, for I very much respected their ability.

The Design Engineers also provided advice if necessary. For example, we worked closely with Stress Office who provided damage limits for us to put into the manuals produced by Customer Support Technical Publications Department, that ensured that damage caused to the fan blades by debris off the runway surface or by bird strikes was repaired safely. These limits were developed as the fleet exposure grew, and all based upon the need to keep the fleet safely flying. We never had a fan failure as a result, and we also reduced the need for so many fan spares.

One of the documents that was so very necessary inside Turbo Union was the FMEA, which helped me to identify how to improve QRA (Quick Reaction Alert) for the aircraft. I therefore looked at preflight requirements. The Overspeed Governor check had been a requirement before every flight during the flight development programme, but maybe would not be so necessary In Service. I asked the Bristol Reliability Engineering Department who had constructed the RB199 FMEA (Failure Modes and Effects Analysis) in the first place, to data mine the flight development records to see if they could support after flight instead of before. They found that after flight would be good, but then surprised me by saying that once a day would also be satisfactory, provided there were no more than seven flights in a day! This was welcomed by the RAF (Squadron Leader Mike Moffett MSc Metallurgy) who had asked me what could be done to improve QRA, and thus get the aircraft airborne more quickly.

I had recruited several young degree qualified engineers to help the RAF with the components that our overseas TU partners designed.

Thus Clive Blackman, Steve Warrin and Trevor Turner joined us, all under the management of Mike Munday and Tony Offer who were my two managers. The results were good in that when we had a thrust reverser deploy in flight on P06, they joined with MTU to sort the problem out with our design and Customer Support colleagues in MTU. Similarly when we had nozzle problems with the air motor that drove the nozzle, and so on. This was a successful partnership in my view, but it appears the IMechE refused to let them become Members because they were not doing design work.

I was given this picture as a Customer Support 1985 leaving present which I still have – it shows MRCA prototype P06, emblems from our RR reps around the stations, and almost invisibly the signatures of many who knew me, including RAF signatures. Note that Tornado won the US Red Flag competition in 1984 with 617 Squadron crews and aircraft.

First RAF Tornado crash

On 27th September 1983, a Tornado aircraft from No 9 Squadron, TWCU (Tornado Weapons Conversion Unit) Honington crashed killing the pilot, at Wolferton, near Sandringham. I visited the site the following day, and found the Rescue Team had already erected a canvas screen around the aircraft which was deeply penetrated into the earth. The screen prevented anybody else from seeing what they were doing in trying to get the pilot out of the aircraft, whom they managed to expose while I was there. I was there to see what help we Rolls Royce people (on behalf of Turbo Union) could give to the AAIB/RAF investigation.

CHAPTER 4

SINGLE CRYSTAL TURBINE BLADES

In 1985 I applied for, was interviewed, and became the first (inaugural) RB199 Assistant Chief Engineer and I was then employed on Tornado engines and was told that I was responsible for ensuring the success of Rolls Royce's (and Europe's) first production single crystal turbine blade, that the Bristol Director of Engineering launched when I joined his team – I had never been involved with turbine blades at all before then, but I was to be Project Manager, not designer! The blade had previously been an Advanced Engineering XG20 responsible item, and it had cracked at very young hours, but the Director Tony Woodings (ex Derby Chief Engineer RB211) said that was enough to launch it for full scale production and he ran the first meeting where he delegated the Design Office to identify all of the problems experienced with the blade so far, and to provide solutions. My part in that blade assessment and design programme was to ask Neil Evans (the Design Office Turbine Blade Specialist) to discuss with me my view about an aerofoil crack that I thought had been caused by a big twist, but which he had not so identified. He exclaimed "it's the shroud angle!", and he worked out what angle was needed and made sure it was on the drawings. That shroud angle is now standard on all Rolls Royce engines, after a deep investigation inquiry (an audit) into the logic by the Derby Technical Director and his Derby specialists.

The Bristol single crystal design team were thus audited, and I contributed to it, by arguing that the new shroud angle would not cause the blade to slide vibrate against the blade alongside, and would hold it steady. The original shroud angle prevented sliding, but with a a very big twist to the blade causing the big crack we discovered. It seems to me that the early non single crystal blade may well have had the same twist problem, but this was never explored it seems. The early non single crystal blade failures left the blade burnt at half height, with the top half disappeared, but shroud angle should have been considered as a solution option back then too. I asked Frank Litchfield in the Stress Office to construct a good Finite Element Model of the single crystal blade, so that we could better understand the effects of what we might later find during testing.

I later worked with the RR Derby Foundry Manager, Sue Lyons, where the single crystal (body centred cubic structure) was being cast, to find that the scrap rate was very high, and she said she could identify what characteristics those scrap blades had. She identified which limits on material laboratory limits had to be amended to reduce the scrap rate, and I ensured that priority was given in the Bristol programme to machining and testing these previously rejectable blades – see ISBN 0-470-01821-6 “Crystals and Crystal Structures” for general info. The rejected blades had been kept by her (hooray for her), so it was just a matter of launching their machining and testing, which Development Department did on my demand. The tests were successful and the scrap rate, and therefore the cost to the Customer, was reduced in a major way. The scrap rate reduction also saved Rolls Royce from having to buy many more new furnaces to produce the required high rate of output needed – not just for the RB199 either, for the big civil engines were now going for single crystal too.

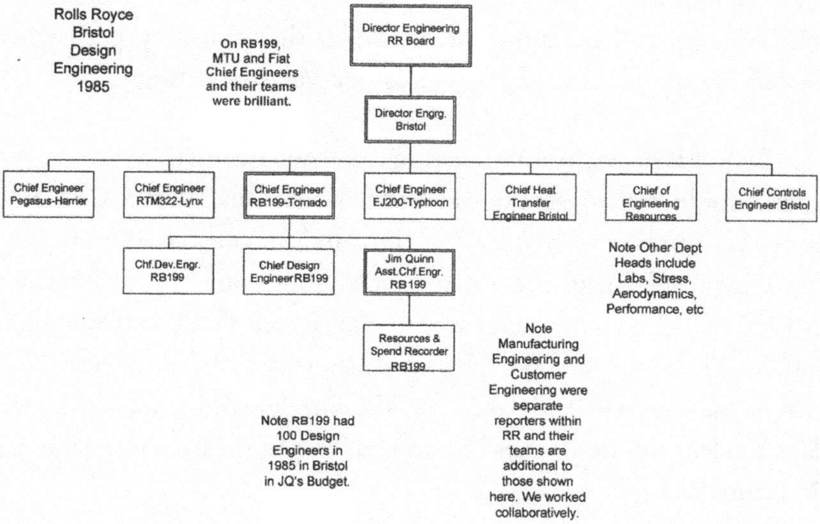

I must mention here that the Machining Department at RR Coventry produced some amazing technology to Electro Chemically Machine very small holes in the blades – at a high rate of throughput too. Holes were first of all being machined along curved edges of the blade at right angles to the surface, and later (to cope with Saudi Arabian atmospheric sand) at difficult angles to the surface of the blade. It became Rolls Royce's (and Europe's) first successful production single crystal turbine blade design and make programme with RR Bristol, RR Derby and RR Coventry all involved.

By this time I had realised how important Manufacturing Engineering were in all of our work. We Design Engineers might like to introduce complex shapes and new technologies, but none of them would be possible without Manufacturing Engineers. Look at the obvious big fan blades on Trent engines today, and you will see how complex these aerodynamic shapes can become with 4D computer technology design tools (4D aerodynamics tools for the rotational shape, and 4D stress tools for strength, impact, vibration too – 4D because it is 3D x y z plus the rotational axis) – but just imagine how to make them for they are hollow multilayer Titanium Alloy. Manufacturing excellence is the past, present and future for Rolls Royce. I spoke to Director Bristol John Wragg, and he launched me down that thinking path when I started the single crystal blade – another Manufacturing lead.

The Bristol Director of Engineering Tony Woodings asked me to write a paper for him to present to the International Gas Turbine Conference in Tokyo. I presented my draft to him, and he said "good, now you go and present it!" I thus presented a paper about our single crystal work in 1987 in Tokyo – my paper was media cleared by Rolls Royce seniors and MoD. It was a good job my paper was to be presented on the Friday, for I travelled to Tokyo on the Sunday and it took me several days to clear my headaches due to the long flight over Russia and the time zone change.

As ACE (Assistant Chief Engineer), I visited Riyadh in Saudi Arabia to discuss the HP turbine blade failure problem with the responsible Saudi Prince and his team, which was due to much gypsum powder in the atmosphere at Tornado's low operating height, and for which I proposed our Bristol solution. The gypsum powder blocked the cooling holes thus leading to very high metal temperatures and failure of the HP Turbine blades at short life as a result. Neil Evans received a payout of £2000 from Rolls Royce for his patent to help the Saudi Arabian Tornado engine blades leading edge. The modifications to the Single Crystal Blade were successfully demonstrated during gypsum sand testing on our test bed at Bristol, and were thus proposed to Saudi Arabia.

The team had made some modified (Saudi Arabian standard) examples to test them alongside our other RAF standard single crystal blades, so that we could see the benefits on a Bristol gypsum test more positively compared with the standard blade. I was gone from the ACE role by the time Neil told me of the £2000 when I met him in the car park. I thought it was too little, for the Saudi Arabian's continued with Tornado purchase as a result of his design. Civilian airliners fly much higher and are not significantly affected by the low altitude gypsum, so the Rolls Royce Board level Engineering Director at Derby did not know of this problem before Tornado entered Saudi Arabian airspace, and the movement of RAF aircraft to Saudi Arabia was not known to us in Bristol, until they were actually on their way. It was a secret kept from us! The Saudi Arabians told us in Riyadh when I went there, that they knew where to put Tornado for their first Operations because they had had similar turbine blade problems with their USA F15 purchase at that airfield.

During all of that time as ACE, I obtained the number of people working on the project in each department from their Departmental heads who had estimated the numbers needed based on the projects Problem Sheet needs. I presented that to Roger Symon the Chief of Engineering Resources, who gathered the numbers of engineers required in total for all the projects at Bristol. At his meeting we were told what the totals of all the projects were, relative to the number of engineering employees, and to adjust our budgets accordingly - downwards of course! I also monitored the actual manpower usage every month relative to that budget, which triggered me if necessary to talk to the Department concerned, when the figures were either over or underspent for that month. This spend plotted against budget came from Rod Stephens, the Resources and Spend Recorder for the RB199. I also ran the weekly A sheet reviews – an A sheet identified the problem that needed to be addressed, so the weekly review first identified whether a new Asheet was acceptable or not, and then reviewed progress on a specific department area, like AGE (Aerospace Ground Equipment) design in the Control System Department.

A Chief Engineer is not necessarily the man or woman who knows how to design an object, nor how to analyse its aerodynamics or stresses, nor its control system, but is there with a wide engineering background to question and suggest changes to designs or analysis. In fact most Engineers do not understand control systems at all. But he or she does monitor, and therefore control, the future priorities and spend on the job's progress in man-hour terms against an agreed budget. The job, I later learnt, was for the Chief Engineer to prioritise and challenge the team in sorting problems out, and using his or her own background to help push the team into solutions. Later, the single crystal blade was already nearly a complete success when a small non-aerofoil failure occurred just before design aim operating hours. A modification was proposed by Stress Office to relieve the sharp corner at the rear of the blade root where the failure occurred, by a small undercut to simulate a rounded corner. The production version would have been a rounded corner, not needing an undercut. This seemed to me to be a quite reasonable solution and able to finally achieve the design aim hours. This was the easy solution to the problem because the deeper turbine disc temperature problem would probably have taken too long and with too much of a design change to cure.

I asked Customer Support to provide me with a Weibull (a log log function) for the failure that had occurred to the blade, which proved it was nearly at design aim hours. I thought it could already be used by our RAF, GAF, GNY and IAF Customers with a probability of failure at just under design aim hours thus providing them with a blade, because it had already achieved several times more life than those in Service at the time. A blade is going to fail at some point in time anyway, so this was only a little short of desired performance even if left as the design was when the failure occurred, but I am sure the new rounded corner would have made sure it was at or beyond design aim hours. I left the job at that point in time to become Chief of Technology Military Controls, when the Bristol Director of Engineering asked "would you like to work in Control Systems" and I said "yes please". The blade was/is a success in production because an RAF Tornado visited Bristol in a congratulatory overfly, with the design aim life achieved by the RAF, for they had really pushed its flying hours to so achieve, with the rounded corner blade. The RAF were also interested in predicting the future spares requirement of course. It was the Design Engineer's expertise that achieved the design aim, and I was pleased to have contributed to that.

I also took over the Advanced Engineering XG20 programme, which was to identify ways of increasing the RB199 thrust by 20%. This was mainly for potential marketing into other aircraft, for Tornado had been already matched with its engine requirement and did not want more thrust. I discovered that the programme was approaching its funding limit, so I spoke to MoDUK who were largely funding the programme, and we agreed to stop further work. Thus I instructed the factory workshops (through the Development Department) to stop as much as they could while transferring the single crystal blade laboratory analysis, brush seals development, and the new 62B fan to the main RB199 programme. The 62B fan eventually became the RB199 Mk105 fan for the ECR Tornado, but this was not for me, but for the Chief Engineer David Pratley to lead the development. Geoff Edwards took over as Chief Engineer RB199 from David for a short time, and the 62B continued further under his replacement David Marsh, who I helped one day with my suggestions about shot peening causing some cracking of the new fan blades through shot-peened compressive stresses being surrounded by a region of tensile stresses.

I visited MoD Rome with, and to help, Lucas (York Road, Birmingham, UK) persuade Italia to adopt our UK Tornado F3 engine DECU electronics. We were in competition with BGT's electronics in the Deutsch Tornado ECR aircraft with its Mk105 engine. Italia, later, did use several of the UK's RAF Tornado F3's with their Lucas electronics. Below is the usual schematic of the DECU (Digital Engine Control Unit): CWP and CMP are the Cockpit Warning Panel and Central Maintenance Panel respectively. Dry and Reheat captions are of dual Self Monitoring CPU. Note that the digital processing has to be surrounded by analogue electronics to process the Inputs, and to provide the big amps current that the Actuators need.

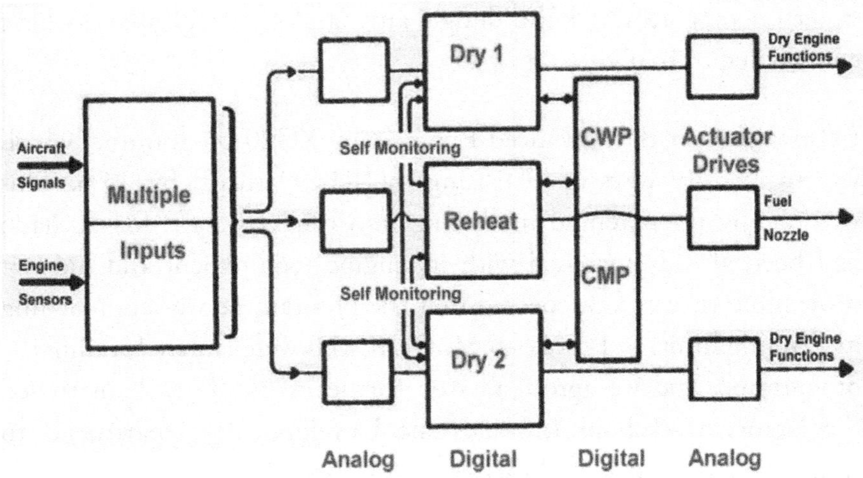

Managers and Specialists

1. A Specialist is always right. A Manager never is in a Specialism! Specialists are highly skilled technical people who sometimes do not make good managers. Specialists should exist up to the highest salary. Specialists are not just a function of the lower orders. Even good Managers might be paid less than their Internationally recognised Specialists who so qualify.

2. Managers should have a clean desk policy. They should have a table around which people can sit and on which they can place their papers for easy access and presentation during a meeting.

3. Managers should facilitate their staff, listening to them to understand their point of view. Managers never talk down to people. They lead by example. They provide encouragement. They make the job fun.

4. Managers should obtain consensus. Engineering is not effective when run on the majority vote principle. Consensus is everyone

working to the same end, but with each of their specialisms having a recognised voice

5. Managers should recognise that the prime attribute of any intelligent conversation, is to pitch it at a level that the whole team understand, not just the very clever few.

6. Managers recognise that speech can often become jumbled, while documents are more easily structured, constructed over time, and thought processes easier to unravel when thinking quietly.

7. Managers never set out to confuse others - they recognise how important it is to retain the trust of their staff.

8. Managers recognise that people can surprise you. They recognise that biology has done some amazing things, and they respect everyone's point of view. People often have a surprising depth of perception.

9. It is not often appreciated that thinking takes time. That interrupting without thinking is not a good thing. Some people think deeply, then speak.

10. When people have to work to earn money, they will sometimes be scared to speak out for fear of being sacked. Yet the hidden emotion is damaging to the team in the end. The culture has to become one where everybody has an important contribution to make.

11. Managers will recognise that sometimes it is better to sit quiet and listen, and thus to let others take centre stage in a protective, rather than aggressive, environment.

12. I work for Managers, not Specialists.

CHAPTER 5

CONTROL SYSTEM DESIGN

One of the jobs that I had to do as Chief of Technology Military Controls, was to create top boss comments upon 80 personal appraisals that each individual would see. The diagram below shows my good managers for those 80 Engineers: Barry Miles, Colin Ford, Geoff Harman, Gordon Toyne, Jerry Croome, Mike Pope, and Tony Northcott – the Project Controls Engineers had no staff. This took a great deal of time, because I was concerned about their attitude to the appraisal and I wanted their morale to be high. I did not criticise anybody, because that person's immediate manager who reported to me had already done that, but I launched training courses for individuals where I thought it appropriate to help, and said so. Where a training course was not identified by me, I said I supported the review. I was told that I was unusual in that other Departments Managers only gave training courses to those who had done well, but I think that everybody needs help at sometime to see what the business was about, and giving to several helped the message to spread across to all.

It was during this time that Rolls Royce Bristol introduced the concept of a Specialist separate from a Manager. Previously, a person was upgraded to be a Manager even though they were just doing a Specialist job, and this led to some confusing management! The separate concept allowed the World Class Specialist to earn more than his Manager, which I thought was a very good thing for Rolls Royce, because Management then became Management. It seems to me that Derby Directors had not learnt this lesson. I was a Control Systems Manager, not a Specialist.

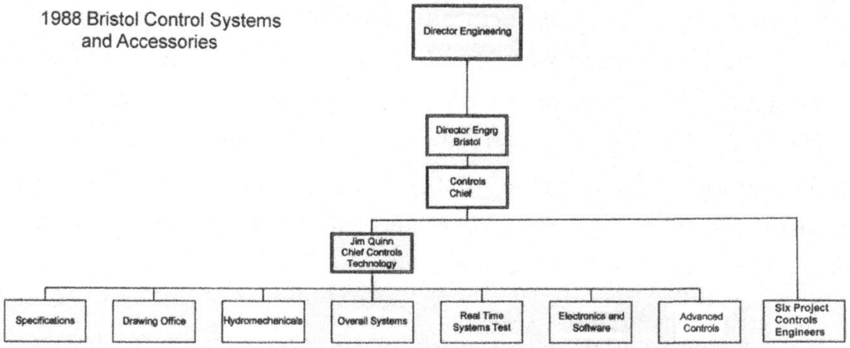

I started the Chief of Technology Military Controls job just as the AMECS (Advanced Military Engine Control System) programme supported by MoDUK was finishing, but you will find part of its work in an interesting book about computer parallel processing ref ISBN 3-540-19747-8 "Parallel Processing for Jet Engine Control". This book shows some of the quality of creative thinking that "my new team" were involved in and who they consulted with. I did meet Professor Peter Fleming just at the last meeting in Bristol, and the author, Haydn Thompson, was there too. This work is unclassified of course. One of the infuriating aspects of Derby was that the Technical Director reviewed the AMECS work before I arrived in 1988, and then prohibited the team from talking to MTU about it to help Eurojet make the control for the Eurofighter Typhoon. He obviously did not understand collaboration, for that meant that EJ200 development was going to cost more as MTU explored the same logic as AMECS had, and for which Bristol and MoDUK (who largely funded the work) would have receive good plaudits! He was also not worried about ignoring all the work that the team had done, effectively just dumping it, with all the team upset that led into.

Procedure was all important to me and one of my responsibilities was to ensure that procedures in the Department were up to date and followed. I thus made sure there was a rota of procedure number against appropriate personnel and monitored that this was performed on time. I was able to audit them too, but there were many and I never got around to looking at them all. However, this was interesting for it told me how things were to be done, and helped me understand the organisation – my next step was to see how the Adour FADEC was created, and I started to take an interest in that, but I was moved from the job to be the first (inaugural) Head of Advanced Controls across the whole of RR Aerospace, and thus to start again with Derby demands and to construct an organisation.

However, I also managed a team of direct reporters to design an electronic box to support the US Navy's Goshawk. We achieved on time. I was also negotiating, with Dowty fuel systems, a FADEC for the Adour engine for the first time, and they set up this demonstration for me, wearing protective glasses and holding the throttle. It worked well of course at test bed sea level. FADEC means Full Authority Digital Engine Control, a part of all modern engines today, which receives sensor values from around the engine and aircraft, and drives the fuel and nozzle actuators. The Digital part relates to the software chips inside the electronics, but you need to surround that with analogue electronics to drive the high current actuators.

1989 - Driving the new digitally controlled Adour

One of the other jobs I was asked to do by the Rolls Royce top Derby Engineering Director, Stuart Miller, was to visit SNECMA (Safran) in Paris, and compare the Typhoon control system design with the Rafale engine control system design. This I did using a published Typhoon Engine Controls Paper, but I was not talking to a SNECMA Controls Engineer. He was something to do with Contracts, not at all Engineering and I told Stuart that he was on the phone all the time to get information in response to my questions. I was therefore not in a challenging situation. But Stuart was apparently pleased with my report for he said "good" as I finished my talk with him.

I discovered that the Pegasus engine had been updated to a digital electronic controller, but with fuel cooling which seemed to me to be entirely unnecessary, as the following chart depicts, for the Harrier has no supersonic capability, so those intake temperatures should surely have been OK for the electronics without cooling. I also heard that the flexible mounts on which the electronic box sat had been bathed in kerosene at maintenance times (and up the unfortunate technicians sleeves!) and the vibration levels went very high for the mounts then slid on the surfaces rather than bounced. In retrospect, even if the box temperatures were thought to be too high, I would suggest that the box design be floated on aluminium blocks which would keep the electronics cool for thermal inertia reasons. I also noted that the Pilot's CAU (Cold Air Unit) was mounted very close to the engine and could have been used instead.

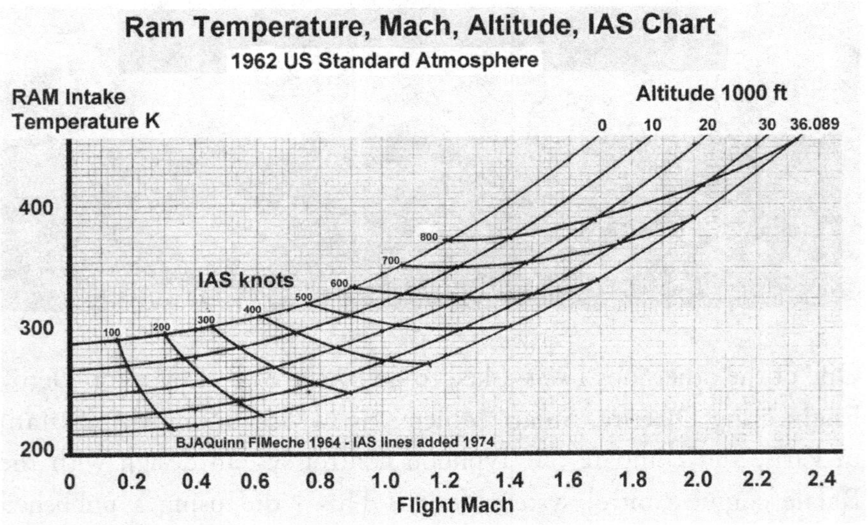

In 1988, a Rolls Royce Director, John Ferrie, from outside Engineering asked us in a meeting in the Sales Lounge at Bristol, to work out how to improve our efficiency – I claimed that was the wrong thing to do, for we were already working at high efficiency and not much gain would be had by increasing it, so we better needed to do more things – using our technology, but not just aerospace. Rolls Royce subsequently diversified, so I won that one!

A new man was appointed Rolls Royce Bristol Engineering Director, and he launched a programme to investigate a Viper engine compressor. I said to him that my Controls Systems Department should have been consulted on it, but he said he was worried that we would tell the Americans about the work. I said "no, of course we would not" and I returned to my office to ask my Sam Shutler to investigate – after all, we design our own control systems functions in Bristol, and then instruct Suppliers, not the Suppliers telling us what to do. Sam came back to me and said that the compressor people had already involved Moog in the control loop. Thus I wondered just what this new Derby man thought he was doing, for Moog was an American Company and word was bound to get out to Rolls Royce's USA competitors – the very concern he had about us obviously not appreciated in his own naive construction of a control system!

I became inaugural Head of Advanced Controls in 1990, and was given responsibility for all Rolls Royce advanced aerospace control systems technology. I was asked by my boss, Tony Roulstone, to launch the HISSC (High Integrity Software and Systems Centre), and to identify the team which I did, but in Derby where unfortunately the people I recruited were not so knowledgable as in Bristol – I now wish I had launched it in Bristol instead. I was interested in this because it would evaluate and evolve the tools which were to be used procedurally in the creation of control systems – I was thinking about the creation of the Adour FADEC at the time which would have benefited the HISSC for it was using tools that were evolving from the very advanced Eurofighter Consortium with its Defence mandated set of advanced computer tools. Simulink would have been one of those tools, and it was just beginning to be used in Bristol. Unfortunately I was moved to auditing before I could develop the HISSC.

I was also asked to think about a "Generic FADEC" presumably to see if there was some way that costs of civil engine controls could be reduced by having just one control system for all their big engines, but I did not realise this until after I left the job. I had obviously quickly dismissed the idea of fitting reheat functionality to civil engines, for it takes up half of the Tornado Engine Control System, but so wide was the directive! The Adour FADEC was not reheated, and I started thinking that the directive meant fitting that control system to all engines, and I was stumbling for a while trying to work out what to do. I realised after a while that the Hawk aircraft in which that Adour FADEC would have fitted, had nothing like the same operational requirements as their big civil engines, but we first started talking to Smith's Engine Controls about "generics" before we had got our own requirements clear (Smith's had made the Pegasus engine FADEC that was later fitted to the Harrier aircraft, and again was not reheated and had totally different operational requirements too).

I came to the conclusion that "no" it was not possible for many years, because while the control functions did not change much between the civil engines, the demands by the different airframers for more information from the engine and its systems was considerable and increasing, for example data-buses (as in our RB199 powered EAP – see ISBN 1-86058-289-3 "Aircraft Systems" and "EAP The Experimental Aircraft Programme" ISBN 978-0-9573755-4-3) and all electric aircraft were, in 1992, being conceived by airframers for their long term strategy. This made redundant what we had at the time, and electronics was certainly still evolving in miniaturisation, which led to the chip manufacturers cancelling further supplies of a particular CPU. We had that happen on the engine DECU for the Tornado F3 where Lucas had to redesign their electronics to change to a Motorola CPU. For EAP, BAe contracted Bristol to design the databus interface for the DECU, which was successful.

It is now probable that stability has evolved in such airframe/engine required designs to make it possible to use one FADEC type on several different engine types now. However, the Fuel Metering Unit and Fuel Pump could not be the same for both big and small engines, because of too big a turn-down ratio leading to marginal stability and fuel over-temperature on small engines, and who in 1990 knew how big the big engines would become anyway? Between the 220 kN of the RB211-524G on Boeing 747's in 1990, and a never imagined 430 kN Trent XWB maximum engine thrust in 2017 – and the BR700 has only 68 kN (about 15,300 lb f) thrust, so could not possibly use a

Trent Fuel System. Many sensors and some actuators might possibly be used on several different engine types of course, but the harnesses would be new, each possibly needing to cope with more and more EMI (ElectroMagnetic Interference).

I also personally started to investigate the differences between RR Derby and RR Bristol's interpretation of EMI (Electro Magnetic Interference), and the differences between their solutions, but I was publicly moved into my next job, auditing, before I completed that, which I very much wanted to do, for I was told that civil engines needed a higher EMI clearance than do military engines, and that was interesting to me. I was more into doing such things than watching others doing something unknown during very brief audits. I was aware of the Luftwaffe Tornado that had crashed on 6th July 1984 as a result of flying too close to a very powerful radio transmitter near Munchen, and thus saw protection against such high output transmitters as essential, for both Military and Civil aircraft, with an obvious need to avoid such transmitters too. This transmitter (Free Europe Radio) has now been abandoned, but I think we need it back to transmit accurate data to the Russian people after Putin's invasion of Ukraine, which is what it was very powerfully originally being used for.

I went on a Management Course in 1990 during which we were asked to think about our own disciplines, and what we might do to improve them. These styles of management course taught us about Company Accounts and how to interpret them. They also made us take psychometric tests to give scores against leader, moderator, creator, innovator, manager, organiser, evaluator and finisher. All qualities of an individual. We were also given the following ten challenging intrepreneur rules to consider, for we were people who relied upon other parts of the organisation to operate a business! An entrepreneur is somebody who starts up a business and takes a broad view of many disciplines, and thus different from us specialist intrepreneurs operating inside Rolls Royce.

THE INTREPRENEUR'S TEN COMMANDMENTS

1. Come to work each day willing to be fired.
2. Circumvent any orders aimed at stopping your dream.
3. Do any job needed to make your project work, regardless of your job description.
4. Find people to help you.
5. Follow your intuition about the people you choose, and work only with the best.
6. Work underground as long as you can – publicity triggers the corporate immune mechanism.
7. Never bet on a race unless you are running in it.
8. Remember it is easier to ask for forgiveness than for permission.
9. Be true to your goals, but be realistic about the ways to achieve them.
10. Honour your sponsors.

In my discipline as Head of Advanced Control Systems I had had the opportunity for the first time of looking at how RR Derby did things before, at, and after project launch level (my job was pre launch), and I observed that they were launching control system suppliers concept work late, after they had done their own 1.5 to 2 year concept work on performance, stress, aerodynamics, and so on, followed by 3 to 5 years of development. This said to me that Control Systems were bound to be expensive because they had to crush their whole programme into that 3 to 5 years - a 1.5 to 2 year shorter time than RR had. They probably formed two teams – one to give RR something to test their first engines with, and a second to design and make the final version to go on final test and thence to RR engine deliveries to Customers. I think their manufacturing would have to make room amongst their machines and schedules to ask current Customers for delays thus to fit their new engine requirements in. To certify the flight worthy system would require additional test rigs in the Suppliers factories to accumulate the necessary number of hours testing, for that testing would also start late. Thus I thought that as much as £50 million pounds could be saved!

Rolls Royce Engineering Directors were blaming Lucas for high costs, when it was actually them who were causing a big chunk of it, by delaying Suppliers launch, I said.

I am convinced that my proposal for an early partnership with Suppliers is/was essential. My Controls Boss, Tony Roulstone, immediately after that £50 million saving idea, launched an investigation about how Suppliers might be involved earlier. So he and the Buying Department (the people who laid the contract on a proposal from an outside company) later launched a collaboration agreement with Smiths Engine Controls people who had created the Bristol Pegasus DECU – "RoSec" was formed. Rolls Royce later formed a strategic partnership with Goodrich with a bigger investment in the Control's advanced concept stage of a project, but even better, RR now have their own Controls Systems Company able to start a System Concept at the same time as an engine concept started - so I won in the end!

In 1991, Rolls Royce decided to run a redundancy programme, and all 3 of my aged Managers in Bristol left, even though I had asked my boss, Tony Roulstone, to delay their leaving timescale - by the Company agreeing that their terms would still be met later. The Company apparently did nothing, and my Managers took advantage of the redundancy payments and quickly left. One of them, Geoff Harman, said he would not have minded working longer, but said the payment was too attractive and may not be made available again. None of us knew that RR would extend their redundancy programme for many years afterwards, so it would have been possible to have them leave at different individual times instead of all together, and save me much stress.

I then worked very hard to try and do what they did, while also interviewing several and thus choosing one replacement Manager at Bristol, Dave Rosser. I also recruited Mike Gainford as my Advanced Controls Manager at Derby, a post not previously filled for there had been no advanced engineering work in Derby Controls Department before I became Head of Advanced Controls. Mike's job was to recruit Engineers to work with the Advanced Engineering Design Office in Derby. I became over-tired, and later, in May 1992, I went to sleep at the wheel of my car on the M4 near Swindon and crashed heavily, breaking my left arm and hurting my head psychologically.

For successful team problem solving, the psychologist Belbin identified eight roles as follows (in the order in which I scored on the psychometric test on the 1990 Management Course!):

Leader, creator, innovator, organiser, finisher, moderator, evaluator, manager.

A leader identifies the task to be done and motivates others to do it.

A creator generates original ideas to approach the task, and produces several alternatives.

An innovator finds resources in the environment and identifies opportunities to use them.

An organiser develops plans to use time, money and physical resources and to make ideas work.

A finisher follows through on plans and attends to details in completing tasks.

A moderator identifies people to work on the tasks and involves them in it.

An evaluator analyses situations and makes judgements about alternatives, plans and results.

A manager develops plans to utilise people and to resolve conflicts between them.

This Belbin assessment seems about right, but I would put Organiser above Innovator.

CHAPTER 6

AUDITING

On 6th May 1992 I became the first (inaugural) Control Systems Auditor, but I received no training for the job. A few weeks earlier, it had turned out that there was no interview for this new job either, and I was merely asked if I "wanted to work with the best?" I replied "yes, of course" (thinking that I already was), and in retrospect that was the interview over! It was a couple of months before I was told to move to that job with no further information nor interview about the job at all! I was happily engaged with developing my wanted Advanced Controls job, and I did not want any other job, but this new one (it had never existed before so I had no idea what they were getting me into) was made public that I was doing auditing before I was even told privately – and there was no consideration or discussion of how or what I wanted to do for my own future plans! I accepted the job in the end and thought I had better get on with it. I was eventually named Chief Engineer Control Systems Audit.

The first job I had to do was go to Lucas to sort out why the new Trent engine on test was "nodding" - it was cycling up and down on thrust and rpm enough to be of concern. I discovered that RR Derby were not simulating the engine control system functions at all, and so actioned them to construct a simulation so that they could better determine the functions that caused the nodding. As it turned out, the Lucas fuel metering unit casing was too thin and flexed as its control function cycled. The casing was thickened and the nodding disappeared. Thus the reason for testing engines on test beds before release to the Customer of course – this was not a case of Lucas doing things incorrectly, but of them trying to reduce weight just as RR wanted, and is one of the hazards of pushing technology as far as possible – a normal fact of life and nothing to blame Lucas for. However, they might have been able to find this on their test rigs if they had had the time, but RR were not giving them the time – I hark back to my £50 million whistle blowing paragraph earlier.

Control System Auditing as a job to me was really still 3 or 4 years away, for I needed to get my feet wet in Control System's Design, of which I had little experience, and I thought I needed all of that time to learn first. However, I started straight away. I also needed to understand how civil engines were qualified – it was baffling what rules were to be followed and thus how it was done – I had never had anything to do with the CAA before, and had no knowledge of how one qualifies a civil engine, but Derby seemed to insist that there was something wrong with Bristol, simply because I (we) did not understand how different Derby Management was. I was used to working with airframers not Certification. I did not know the overall Departmental organisation at Derby either, so had no idea who to talk to about qualification of a whole engine, while in Bristol I knew many People by Name quite well. Even the Control's organisation at Derby was different from Bristol's. I should not have been employed doing that auditing job for another few years, for while I knew about Control System usage (like I knew the details of the functions generated by the electronic box that was the MECU or DECU because I had asked for some of them when I was in Performance Department), I had no idea how to design or implement a control system working from inside Rolls Royce. Sir Stanley Hooker left Derby as a result of problems with his boss there, but fortunately he was able to join Bristol Aero Engines, a very much happier organisation. I had nowhere to go that wanted my particular expertise, for Rolls Royce was now the only employer in aerospace gas turbine performance in the UK.

My auditing job was a job headed by the new Technical Director. I was told by my new Boss (Deputy Technical Director) just to listen to what I was being told by the team I was auditing (unfortunately, I am not a good listener of long wordy descriptions because I have a hearing problem. I can read and think much better than I can listen). I must have participated in a fair number of audits with the Technical Director heading the audit team, but he never introduced us to the team being audited nor to each other, and never told us why we were auditing that particular engine at that particular point in the engine programme either. At Concept the project is all Technical thinking (what might work, what options are available, what plans needed for new aerodynamics testing, and so on) while, near Certification or Clearance for Flight, I think the main concern is keeping the Project on time, since all the Technical Decisions had already been made. I had been thinking that a Technical Director therefore should not be auditing a Certification phase engine. However, and in contrast to what was actually happening, he or she should review Departmental tools and procedures especially.

However, an audit that I conducted very successfully in January 1994 was a detailed 2 day Launch Review (I prefer the word review) of the BR700 engine control system concept which I enjoyed in Munchen, where the good individuals in the audited team presented their technology on projected slides with many good diagrams which I happily enjoyed reading and discussing. They treated me well, with their careful presentation which was so often missing on other audits. The BR700 team had obviously had enough time to prepare, and I wonder if those other audits previously had been given enough time to prepare – it was rare that I had notice of an audit more than a few days previously. There was no preplanned schedule of audit dates that I saw. Anyway, on this BR700 review, I was the Lead Auditor with Alec

Collins quietly alongside me, and a senior Deutsch engineer alongside him. Alec became quite excited about having aircraft nose pressure input to the engine control system, just as we do on Tornado with the TTU (Triplex Transducer Unit) feeding both engines.

Gulfstream had told the BR700 Controls Team that it was a better idea than measuring pressure in each engine's intake because nose pressure improved aircraft control in side winds. This was because engine intake pressure would be blanked (and thus reduced) on one side in side-slip, and thrust on that engine would then be reduced making side-slip worse, through the engine mounted pressure ratio control function. Aircraft nose pressure input to the engine control meant that thrust would be maintained as Gulfstream wanted. I had not realised until this BR700 audit that civil engine controls had an Engine Pressure Ratio function for take-off. I presume they have it this way because thrust is maintained as the engine gets older? I should have had a short burst of training in Derby Performance Office beforehand. On Tornado we have a top temperature control and modulate reheat to control thrust on formation take-offs. Note that Derby's RB211-524G FADEC technology (Boeing 747-400 in British Airways) came from Bristol and Lucas.

I had flown from Bristol or Heathrow to Munchen well over 60 times by the time I left – 62 flights abroad were recorded in my passport between 15th April 1966 (I flew to SNECMA in Paris 7 times before the AFVG was cancelled) and 8th October 1979, for in those days the UK Government had a restriction on the amount of money you could take out of the country, so each trip was recorded in my passport with the amount of money being taken out. I flew to Munchen 25 times before MRCA P01 first flew on 14th August 1974. I purchased Haribo bears in Munchen Flughafen before they were available in the UK, to bring home on numerous occasions for my kids, as they still happily remember!

An audit is about a review of a particular discipline or engine or department. To review objectives and to spot loopholes. To think about how improvements might be made, and by whom. To ensure the team audited have covered all of the options. To thus ensure the highest integrity possible within cost constraints. The Rolls Royce audits that I covered were all about auditing a whole engine each time, and I did 37 of them. I thought about those audits and concluded they needed something extra, for I thought they also ought to be about the specialist technology in each discipline – fan, HP compressor, Combustion Chamber, HP turbine, LP turbine, Control's Software for example. Thus the HP compressor aerodynamics department, for example, would have specialist knowledge of all of their latest technology and computer tools and procedures, and would have addressed each engine as it demanded an HP compressor, getting better as time went past. I thought that auditing technology departments was the best way to challenge the latest technology available (better than engine level audits), and to ensure its competence, its reliability, its cost effectiveness, and its performance. This would be my sort of Technical Director job, because it would ensure the cost effective growth of a strategy for that component in isolation, but necessarily fit for an engine.

Control systems were evaluated at engine level, and they are all similar at systems design level, but software was a learning process and should have been conducted as a technology in its own right. The software was split into two – that designed by the electronics supplier to check their units function, and that by Rolls Royce for the flight software. It was the flight software that should have been checked for its technology level, by the Technical Directors audit team. The technologies of the various control system components like fuel systems, electronic controller design and its electronic components, and other sensors and actuators which would all be developed by the outside supplier, were selected by the Control Systems department against technology offered and costs. I did not know how the pressure ratios of the three fuel pumps between aircraft fuel tank and engine combustion chamber were decided for example – the tank pump, the aircraft pump and the engine pump all to provide fuel at high pressure to the engine combustion chamber fuel jets. To do all of this properly at a delegated Technical Director level, would require design expertise and exposure to various submissions made by each of the suppliers too. I was good at interpreting the behaviour of the RB199 control system design on the engine because of my Performance background, but I had no experience of selecting supplier's or Rolls Royce's designs. I had no knowledge of how different technologies emerged on each of the many components in a control system. And of course, the technology on offer from a supplier was based on their own R&D, not on Rolls Royce's.

After I left Rolls Royce, in November 1997 I paid for my own 4 day training programme (about £1500) and became a Lead Auditor of Quality Management Systems. I learnt what that sort of audit was all about, and was told that ISO 9001 is not a Quality Mark for it merely identifies the relationships between departments in a company and not at all how good that relationship was. We practiced a simple audit. This developed my thinking and helped me to construct the logic outlined above for Rolls Royce Technical Director auditing of engines and their components. This training should have been part of my introduction to Chief Engineer Audit, but the Directors had not thought to train me.

I will be attending the two day December 2022 IMechE Design Review Course for interest and as a focus for discussion. I should have been sent on this Course or something similar at the beginning of my Auditing Career with RR.

CHAPTER 7

HANG GLIDING

In 1978 to 1980 I went hang gliding. Firstly, I went to a film presentation by the Avon Hang Gliding Club where I saw several films of colourful hang gliders pirouetting in the air, and one of them had Fleetwood Mac's Albatross music playing during it. That really turned me on! I liked the colours of the gliders and I bought an old one, a McBroom Lynx prototype which was all I could afford. The other Avon Hang Gliding club members had moved onto more modern wing-like gliders, and I found I could not keep up with them unfortunately.

Nevertheless I did 27 flights, particularly enjoying Bossington Hill near Minehead in Somerset, where I flew prone for 48 minutes on flight 24, performing several 360 degree turns too. The smooth hill slope was facing west directly over the River Severn estuary, and it was quite high, so that the wind came strongly up the slope from the sea, but you could not fly too far rearwards, because of the wind rotor there which could cause you to dive into the ground, hard. I gained several hundred feet in altitude most times, and learnt how important it was to fly prone (on my stomach in the harness), for flying seated meant that I could almost not fly forward in stronger winds for the drag was so high. Prone reduces the drag and gave me more manoeuvrability as a result.

I discovered how bumpy it was to fly over trees at other venues, and certainly that made me more conscious of why Tornado's small wings and sweep had been designed to minimise pilot fatigue over long distances, and thus to operate at low level over such terrain. I shook hands with my friend Paul Millett on 8th March 2002 at the Thanksgiving Service for Wing Commander Roland Beamont CBE DSO DFC DL FRAeS, and at the reception afterwards Paul told me (and Wing Commander Jimmy Dell OBE who we were talking to at the time) that he had been hang gliding too. I had met Jimmy at Warton previously, and he had thought about hang gliding (I told him the roll rate was not good!). Jimmy had given the eulogy having worked with Beamont for many years, and he was appointed OBE in 1966, was awarded the Derry and Richards Award by the Guild of Air Pilots and Navigators, and the Royal Aero Club Silver Medal for his services to test flying. The photo below shows me seated, not prone, at take-off from Castle Morton in the Malvern Hills in Worcestershire – the take-off point suits an Easterly wind there:

CHAPTER 8

MOTORSPORT

I first drove a powered vehicle when I was 12 years or so old – my Dad put me in the driving seat of his single decker bus and I took it around the yard for a short time! My dad also impressed upon me the need to move at a junction, so that I held nobody up – so often queues build up at junctions (especially at roundabouts) because people so pointlessly wait for a huge gap in the traffic, when there are plenty of opportunities before then to escape. And clearing a junction also helps to ensure you are not hit from behind!

Motorsport CV		**Achievement**
Aug80 – Mar82	Tavern MC Ltd	Company Secretary
Jan81	Tavern MC Ltd	The Committee Rose bowl
Mar82 – Mar84	Tavern MC Ltd	Chairman
Jan84	Tavern Organisation Classes and annually thereafter	Supported Idea
Mar86	Tavern MC Ltd	Clubman of the Year

Events Organised -

Date	**Event**	**Role**
10/11 May 80	Moonraker Rally	Passage Control Marshal
14 Sep 80	Tavern Autotest	CoC & Secretary
13/14 Dec 80	Omega Road Rally	Secretary
22 Mar 81	Rothmans National Stages	Results Chief
13 Sep 81	Castrol Autotest	CoC & Secretary
13 Dec 81	Octavian Stages	CoC (Clerk of Course)
6 Mar 82	Rothmans National Stages	Results Chief
7 Mar 82	Rothmans Rallysprint	Results Chief
21 Mar 82	Marby Stages	Results Chief
22/23 May 82	Castrol Moonraker Road Rally	Secretary
7/8 Aug 82	Gridnight Road Rally	Chief Marshal
19 Mar 83	Castrol Stages	Timekeeper & Results Chief
22/23 Oct 83	Orion Road Rally	CoC
2 Nov 83 – Letter of congratulations from Avon & Somerset Chief Constable		
20 Nov 83	RAC Rally, Ashton Court	Competitor Liaison Chief
18 Mar 84	Tavern Stages	CoC
10 Nov 85	Eclipse Chieftain Stages	Results Chief
24 Nov 85	RAC Rally in Forest of Dean	Stage Commander
February 86	Wyedean Stage Rally	Stage Commander
20/21 Sep 86	Orion Road Rally	CoC
4 Oct 87	The Pines Stages	Results Chief
19/20 Sep 87	Orion Road Rally	CoC

17/18 Sep 88	Orion Road Rally	CoC
9 Oct 88	The Pines Stages	Results Chief
14 Oct 89	Great Western Stages	Assistant CoC to Dave Whittock
9/10 Oct 93	Bristowe Road Rally	Chief Marshal

As Clerk of Course I evolved the rules and procedures that were needed to both create and run a Road Rally in accordance with the RACMSA (Motor Sports Association) rules of course. I drafted the Regulations and created the Route and its Timings for an average 30mph on unclassified (60 mph limit) roads – the roads I chose were very narrow and twisty, and sometimes bumpy, so it was rare that anybody arrived early, but clearly the best navigators and drivers on these cumulative Competitive Routes won. I arranged for the Start and Finish venues to be open at the unusual times that we wanted, and I ran the Drivers Briefing before the Event started. I agreed the route with the Avon and Somerset Police and the local RACMSA Rally Liaison Officer. The 30mph roads in towns and villages were to be travelled at 20mph average maximum Non-Competitively – easy in the very early hours of a Sunday morning. Competitors were penalised for arriving at the next Control early – and noise and radar speed traps leading to exclusion if exceeded were often arranged too. We all learnt that we could not judge our speed except by looking at the speedo, for we have been travelling quickly between hedges and the brain adapts itself to relatively high speed that way, so judging 30 mph just by looking through the windscreen would probably have you actually doing 40mph or more!

I drafted the Procedures that the PR teams had to follow and the Letter that they gave to residents along the route. I found a Course Opening Car and crew, and the Chief Marshal normally found a Course Closing Car and crew to pick up all of the marshals road side signs. I drafted the time-cards that the competitors had and which had to be filled in by the marshals as they arrived at each post. I also drafted the Procedures that Marshals had to follow on the night, and for the Competitors too. Navigation was by way of Ordnance Survey 1:50,000 scale map references avoiding identified black spots and thus to define the shortest route between the defined references. I borrowed and expenses paid for the RACMSA Official Timekeeper's many accurate clocks that the marshals would use, and made sure that the Results crew knew the rules too. When I had been Results Chief myself I made sure that Competitors Time Cards were collected appropriately, and delivered to me in the Results Room where my computer helped me to sort them into order as their times accumulated. Half an hour was allowed for Competitors to check their Results, and then we presented the awards which I had also arranged. My motor club friends were many, but I particularly remember Bob Rutherford, Paul Waterton, Hugh Currell, Steve Jones, Martin Jubb, Martin Clark, Russell Boulton, Dave Smailes and Mark Milton (who navigated for me more than anybody else).

After the Rothmans Events, I invited my family to come along and help me with Stage Rally results, and so we had a family occasion doing motorsport on those days. Mary and both of my teenage children came along and my son told me recently that he thoroughly enjoyed doing that – I had programmed my Sinclair Spectrum to sort the cars into fastest order, but needed the family to help feed in the stage times for each car, and any penalties a car crew had suffered. Not just Philip, but the rest of my family enjoyed that for they all happily came along on all the Results events afterwards, and we achieved a good reputation for rapid results too. Procedure and organisation was important to our success. It was also important when running a team as Clerk of Course, which I was also successful at.

I gave up smoking cigarettes in 1980 so that I could afford motor-sport in my ordinary car, and I have never smoked since. I wanted to go rallying for in that sport you compete along narrow and twisty roads against your own ability, while in racing on easy big circuits you are with others who can easily cause you to be involved in an expensive high speed accident, so you do not compete against yourself. You have to finish to win! And in rallying you can use your ordinary car in night rallies, top speed is not so important, whereas you need expensive tuned engines in racing to get anywhere. Also rallying is far more interesting from an organisation point of view for you have to plan a route, talk to the Police about your route, and then the team visited every house along that route to advise about the passage of so many cars at certain times of the night. You do the same organisation for day time stage rallies, and I enjoyed doing that too, but to compete with your own car is more expensive (for you can so much more easily go off-road or roll it). I enjoyed co-driving on stage rallies. I was a good navigator on road rallies too, for I thoroughly enjoyed the maps.

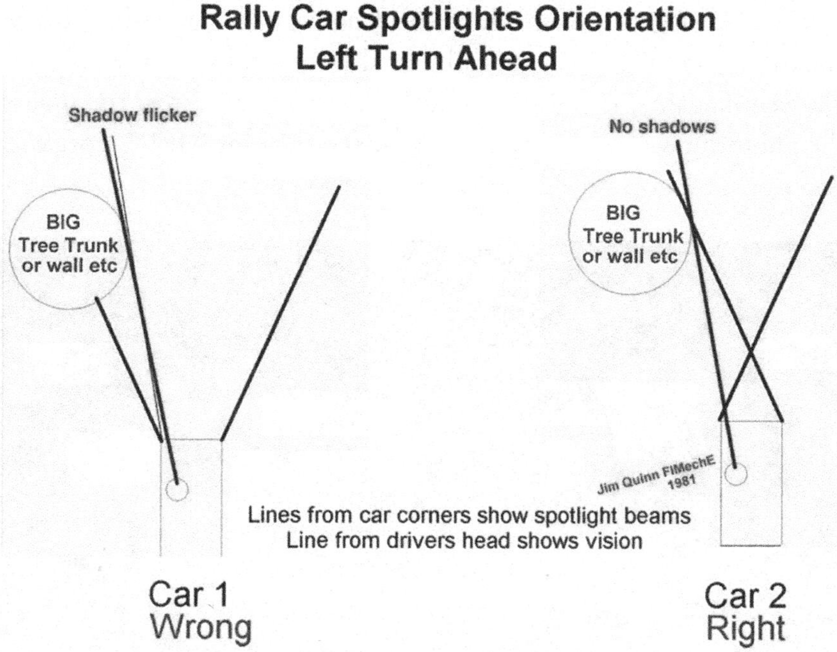

The diagram here shows how to setup your spotlights to avoid shadow flicker at night. Clearly Car 2's crossed spotlights are better to avoid shadow flicker, but many World Rally Competitors have the wrong setup for they are often as Car 1!

I had a Ford Escort Mk1 with a 1500 cc engine in 1982, then a 1960 Austin Healey 3000 in 1987 for the remainder of my time as a competitor. To make the Escort do what I wanted, I uprated the brakes and the suspension, by using Group 1 struts at the front and soft CD6 springs at the rear with metal bushes instead of the rubber ones, and with large diameter Ford Capri drums at the rear. I also fitted a limited slip differential, and a hydraulic handbrake. These proved to be just what I wanted and I achieved some good results on road rallies. I broke the 1500 Escort engine after a few rallies for some reason, so I purchased a second hand 1300 engine which I cleaned and tuned with a better camshaft and pistons. In my 1300 I won "Best 1300" on a few occasions, and often came in the 1300's top three with an overall position (among 2 litre machinery) of about 11th or 12th – I have quick reactions and was especially good in fog, pushing myself up to 5th overall against two litre machinery in one event across the moors. The photo shows us (me driving) in Somerset in my Mk1 Escort with plenty of spectators on this competitive section!

Incidentally, have you ever worked out how quick the reactions of pigeons and other formation flying birds are? Measurements have been made of beak pecking reaction time, but this is much slower than wing manoeuvrability, for static food does not need a quick time. I am 1800 mm tall and my hands can react in 0.25 seconds, but speed of the signal through my spinal cord is perhaps 120m/s (about 0.015 seconds for my height), so the delay is mostly in my brain complexity. My brain diameter relative to the pigeon is equivalent to our height difference and complexity is related to volume I guess, with a minimum reaction time defined by neuron transmission rate. In which case pigeon reaction time to their wings is likely to be 0.01 to 0.02 seconds, so no wonder they can fly so closely in formation together!

In rallying, I concentrated on being very quick around corners, and on slow ones particularly you spend more time going around the corner than you do on fast corners. Much time overall can therefore be gained by deliberately pushing slow corners relative to fast corners. 10% improvement on a 20mph corner taking several seconds from braking to accelerating gains more time than 10% improvement on a fast corner taking one second. I achieved good results and actually won one night rally in Wales in 1991 in my 1960 Austin Healey 3000 Mk I, without spotlights and only the relatively dim glow of my 1960 style headlights – with Mark Milton navigating. I finished up 3rd overall in the Welsh Historic Rally Championship that year. I was observed to be the only competitor to negotiate impossible hairpins properly – by hand brake spinning the car in the narrow road hairpin junction, while other competitors travelled further down the first road to find somewhere to 3 point turn, then to travel into the second road! The roads were only just wide enough to fit the length of my car sideways whilst handbrake spinning, and I was very successful doing it on several rallies. I had also practiced going downhill quickly, for most drivers do not do that, so you can gain places that way. I was Course Car in 1991 for the RAC International Historic Rally (I had spotlights for that event) around much of England and Wales. I was given a "Driven with Gusto" medal for my driving on the event, as reported in the "Thoroughbred and Classic Cars" magazine of June 1991, with a picture of my Healey on Plymouth Hoe. The differential broke there after my frequent use of the handbrake on such a testing event.

The main difficulty I thought I was going to have was getting through the noise test, because the Healey had a resonator by design, not a silencer, but it proved marginally OK! In the events themselves the problem I had was with the front brakes because the discs were quite thin and quickly glowed red hot causing a lack of braking effort as the hydraulic fluid boiled. I often had to use engine braking instead. The discs were thickened on the Mk II and Mk III Austin Healey 3000. The hydraulic oil boiling slowed me down after several miles into an event, as you might imagine.

Interestingly, I finally discovered the original registration number for my Healey – it was 180CAB, but it was an interesting journey getting there, for the Worcestershire County Council records showed the number was registered to a person, not the car – the entry in their County records was "Austin 2912cc", and nothing about the chassis number nor that it was an Austin Healey! There were very few "Austin 2912cc" so I did not have to search for many original owners. I eventually found the original owner whose wife gave me copies of the original invoice for the car which showed it was indeed 180CAB with chassis number "H-BN7/12248", engine number "29D-RU-H/22183" and it was first registered in 1960 in Worcester. I took it to Dublin for five months in 1995, and had it temporarily registered there – they gave it the number 60D25, because it was the 25th car they had registered with a manufacture date of 1960. It was back to the original UK registration at the end of 1995 when I returned us to the UK.

I was a successful experienced motor sport competitor and organiser, 1980 – 1993. I was Clerk of Course on 11 occasions, Results Chief on 8 occasions, and so was an organiser on a total of 24 occasions. As Clerk of Course (CoC) I was in charge of the whole rally. I planned the route and presented it to the local RACMSA Rally Liaison Officer and also to the Avon and Somerset Police, both of whom advised whether there were any known blackspots along the route. A blackspot is where a member of the Public or a Policeman had complained about previous rallies or non motorsport events, and horse disturbance at night was one particular thing we had to ask the owner to look out for. If they could not move the horse away we had to replan the route to avoid complaints to the Police of course. I appointed a Chief Marshal to organise the PRing by groups, and I took the overview of PR completion, to ensure we had missed nobody. PR is Public Relations. We kept a record of who we had PR'd by asking them to sign our petition sheet.

As Results Chief on 8 occasions, I worked with a Professional Computer team on 3 early rallies, and used my organisational skills to arrange for collection of the competitors times by my own organised group of motorcyclists, from marshals in various places around the County. On the remaining 5 rallies I used my Sinclair Spectrum computer to compute and display the finishing order of competitors on each test and then overall, using machine code level Z80 Assembler. I received a congratulations letter from The Avon and Somerset Chief Constable in 1983 for the good organisation of the October 1983 Orion Road Rally across Exmoor, for I was CoC on the Rally – a good organiser.

MOTORCYCLING

I was given a Triumph Tiger Cub T20 on my 16th birthday – what a surprise and a new one too, so It must have cost about £200. An enormous sum of money for my parents to find, for we lived in a council house and were quite poor, so a bank loan was the necessary order from my Dad apparently (I knew nothing about that at the time). Nothing of that value had ever come into the house before! I loved the reverberating sound it made when accelerating over a railway bridge in Erith with high walls either side! I passed my driving test first time a couple of months later (I had ridden bicycles for years so knew about road craft already), and then it was four years before I bought a car instead, and I passed that driving test first time too.

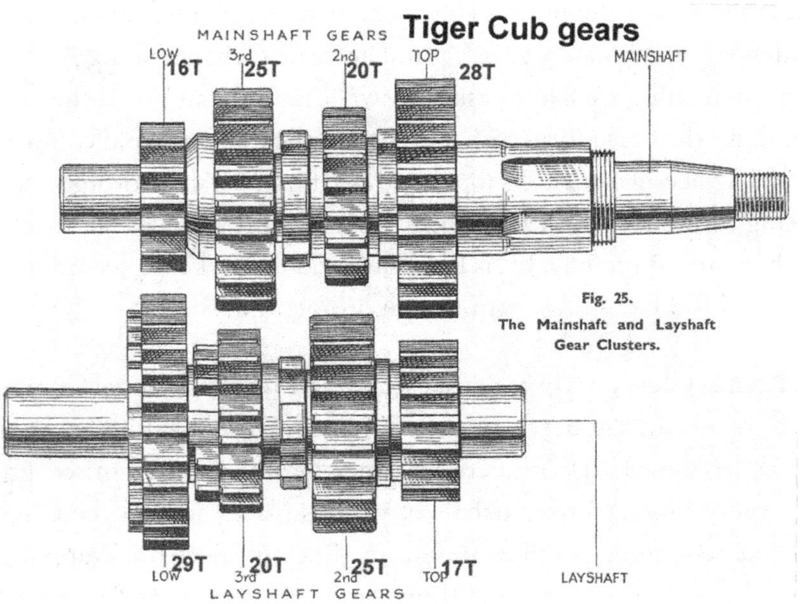

It was as a result of having the Tiger Cub that I learnt about gearboxes – that the sum of the gear tooth count on both mainshaft and layshaft were always the same no matter what gear you looked at – as you can see in the diagram here. Much later, in 1996, I bought myself a trials Tiger Cub, and found it jumped out of gear. I stripped the gear box to find that somebody had put the wrong tooth count gear into it – that tooth count I had learnt about when I was sixteen came in very handy then, for I was able to easily identify which gear I needed to change!

In 1960 I bought a Stanley Schofield 33 rpm "Sound Story" vinyl record about the 1958 Isle of Man TT, which had some wonderful exhaust sounds on it, but I could not afford to visit there myself until 1995! Nevertheless, as a teenager, I often went to Brands Hatch in Kent to watch the bikes and well remember seeing Derek Minter there on his Manx Norton, sounding wonderful, but spectacularly accelerating hard while banked over around Clearways – I had been told not to do that for I might come off, but I ignored that afterwards! I also visited Castle Combe in Wiltshire to watch the bikes, and remember clearly hearing Mike "the Bike" Hailwood on the 250cc Honda six accelerating hard on the back straight – and I was on the finish line, but could hear the exhaust very clearly!

Terry Cox and I in 1963 had 250cc Ariel two stroke twins (very smokey for 10 miles after start) during our apprenticeship together, and I used to stay at his parents house in Hungerford sometimes – it was a good way into my journey back to Erith from Bristol. We used to travel along the A4 without speed limits in those days, so 80 mph was common as we rode together. I was once caught by a Police Car going down Shooters Hill in Blackheath, London – he waved me down so I stopped, but he could not stop so easily for he was another 100m further down the road from where I stopped! A small fine resulted, which I did not suffer again for over 20 years. My first time at 100mph was in a Police Car, a Jaguar, in which we apprentices were invited to take a ride in the back seat – it was part of a police driving course I think, and it was along the A4 past RAF Yatesbury which closed in 1969. This was based upon the Police Driving Manual "Roadcraft", which was very informative about how to drive properly – beyond the Advanced Motorists driving standard I think.

I visited the Isle of Man TT Races in 1995, with my son Philip, and we watched the bravery of those riders as they swept past at the top of Bray Hill, to dive down it at such high speed! I thought that the motorcycle riders would have made good Tornado Pilots for they must have the same sort of bravery! I bought a CD of the 1950's and 1960's glorious TT sounds after that, which I still occasionally play, since my 7" EP vinyl was lost some years ago. Below is a photo of me taken by my son on my four cylinder 600cc Honda Hornet in 2004 – amazing that it had 96 bhp at top engine speed compared with John Surtees 65 bhp from his racing 500cc MV Agusta four in 1958! It was amazing to me to enjoyably feel such power on acceleration, when I had so admired the MV racing around the Isle of Man back then. On the Thursday of that TT week in 1995, I went for a walk around the whole TT Course – 37¾ miles from Douglas where I was staying in a B&B, and I had most people that I met sign my notebook to say what time it was and where I was – I started at the local Police Station and finished there too, both times obtaining a signature – I averaged 3½ miles per hour over that long distance in about 11 hours, excluding the half hour coffee stop I had in Ramsey.

Below, I took this photograph of my son with his 100mph 250cc Kart, alongside his friend with his Kart. My son lapped Silverstone short circuit at over 84 mph in his earlier 125cc gearbox kart at a Kart Grand Prix, doing very well thank you. My son is the only person I know who has actually rolled a Kart – he was OK with some scuffing on his helmet, and thank goodness the high back seat of the 250 cc Kart took all of the load though it was badly cracked afterwards.

While my son was Karting, my daughter was diving in the Maldives. When I visited her there, after a few days in Sri Lanka, I scuffed my feet through the sea in the evening and was amazed to see the phosphorescence that I kicked up – tiny flashing blue lights from the crowded organisms in the water! I was pleased that the Muslim community there treated her well too, for she was there alone for several months as an accomplished Kuoni travel representative. She was an accomplished skier too, and a ski resort rep for several whole seasons. She sent me a card on my 65th birthday, in which she asked "did you ever think you would be 65?", and no, I reckon all of us at that age still think we are only 40!

In 1994 I tried horse riding at the local Weston Super Mare riding school, and was given a big horse called Tag for I was 5ft11in tall and weighed 13 stone. He was odd in that he liked to trot at half mile an hour and he had to be pulled along by another horse and rider to get up to a canter! I flopped about on top of him at that speed, but was secure in the stirrups yet I could not make him do it on his own, and it was not so much fun to me, so I gave up after half a dozen lessons.

Later in 1994 I went "bungee" jumping over the River Thames in Battersea from a crane platform 150 foot high. I jumped twice, but the second time I went off the platform backwards so could not see the Thames underneath me until I was upside down – that launch made me slightly nervous, but I was OK on the first jump forwards.

CHAPTER 9

IMECHE

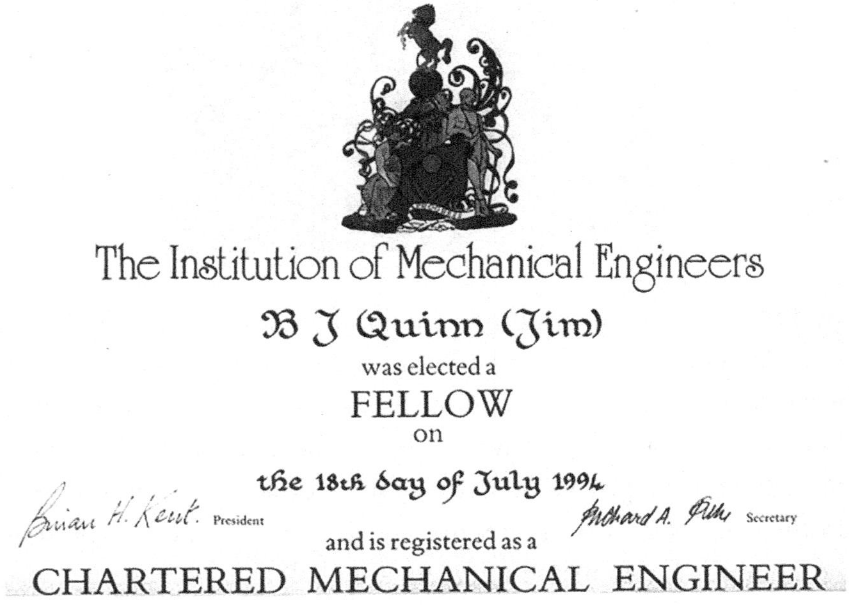
The Institution of Mechanical Engineers

B J Quinn (Jim)

was elected a

FELLOW

on

the 18th day of July 1994

President Secretary

and is registered as a

CHARTERED MECHANICAL ENGINEER

In 1994 I was awarded with the qualification: Fellow of the IMechE (FIMechE). I was already a full Member (MIMechE) having been a Graduate member (GIMechE) since 1967, and I was a CEng (Chartered Engineer) too.

Typically, Fellows are professional engineers working in a senior role with significant autonomy and responsibility and will be able to demonstrate the following qualities:

Essential

Your must demonstrate these Essential characteristics

1) A position of senior responsibility and/or significant autonomy in your particular field
2) Demonstrable Leadership qualities
3) Influencing policy and strategy making decisions in either a technical or business environment
4) A structured approach to Continuous Professional Development
5) The promotion of the Engineering Profession to young Engineers and potential Engineers

Desirable

1) Highly specialist knowledge in a specific area of Engineering
2) Technical or Engineering Resource Management and/or personnel management and development

Optional

1) Responsibility for a budget and the associated risk
2) Application of a significant range of fundamental principles and complex technologies across a wide and often unpredictable variety of contexts
3) Active development and application of new technologies in Engineering and related areas at a senior level

Lecture Organisation

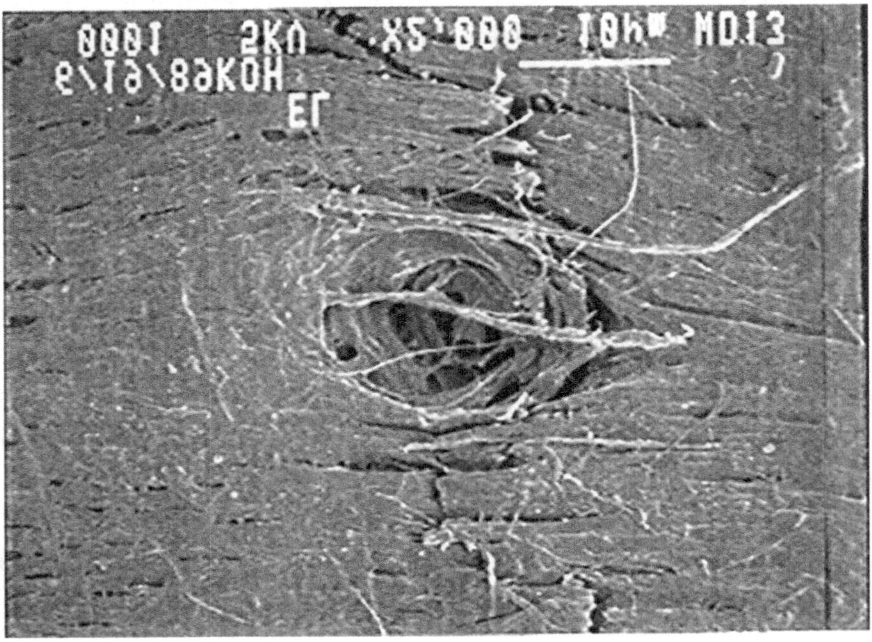

I have worked with the IMechE group in Gloucestershire to arrange two talks and two visits in 2002/3. One of the talks I arranged was titled Biomimetics which is about the abstraction of good design from Nature, and it was given by a Professor from Reading University. It is a multidisciplinary science which looks at engineering solutions in biology, for inspiration and implementation in the man made world. The picture above shows the curved trajectories of chitin fibres around a hole in the exoskeleton on insects. Inside the hole, which acts as a local strain amplifier, is the mechano-sensor used by insects to detect loads in their structure. This concept we were told, was being used to develop integrated strain sensors in composite material structures.

The second IMechE Gloucestershire talk I arranged was by a Professor from Oxford University Chemistry Dept who told us how he used Engineering's Computer Aided Design (CAD) tools to assess the possibility of integrating a selected drug onto a selected protein, both of which are complex geometrical shapes, and once geometrically fitted together sufficiently well (like LEGO bricks linking each other by inserting pegs into holes for recognisable models), they may possibly be able to cure a specific cancer. The Professor's aim was to use this CAD tool to quickly and cheaply find a likely satisfactory drug which would otherwise cost hugely, and take many laboratory hours of trial and error to determine. The proposed drug would still have had to go through all of the appropriate safety and application tests of course, but at much lower expense.

The IMechE Gloucestershire visit I arranged in 2003 was for a group of Engineers to go to Harptree Agricultural College (through my arrangement with the Principle) to be shown around their montage of tools and tractors. As usual for all of our talks, the visit was displayed on a poster and on the local and central IMechE websites to attract customers.

I left the IMechE Gloucester Panel in 2003, but rejoined in 2017 at their AGM, and I became Hon. Treasurer on 13th March 2018. I agreed and ran the 3D Printing talk by Eujin Pei and Giselle Loh of Brunel University on 7th February 2018, and I have contacted and arranged a two hour visit to the Severn Trent Hayden Waste Water Treatment Plant near Cheltenham on 18th April 2018. I have arranged more lectures since: "The Rolls Royce Trent Control System" on 19th September 2018, "The Real World of Fuel Cells" on 18th July 2019 covering hydrogen powered devices and cars, and on 14 October 2020 "The Future of Heating" about introducing hydrogen into our gas network and the boiler modifications needed for it. I email circulate all c.900 Gloucestershire Members with notices of such Lectures and Meetings.

CHAPTER 10

GIFTED CHILDREN

The NATIONAL ASSOCIATION FOR GIFTED CHILDREN (NAGC) had a magazine called CYBERBRYTE to which I freely contributed as a member in 1997.

My letter published in Cyberbrite on 1st December 1997:-

THAT'S A GOOD IDEA: A possible response from friends on the Internet Cyber Cafe – but what should you do then? You might just go and produce a large number to sell, but someone else might copy your idea and take your Sales away from you. There are a number of ways in which your good idea(s) might be protected in law for either your exclusive use or for someone else to use (if they pay you!), but none of them are cheap.

Music, maps and magazine articles might be Copyright. Shapes might be Design Registered. Symbols might be Trade Marked. Inventions might be Patented.

A Patent will cost - £25 to first File it (Form 1/77) with the Patents Office – another £ 130 to Search and – another £70 to Examine possible previous Filings.

The 1977 Patents Act defines that an invention is only patentable if

a) it is a new idea
b) it involves an inventive step
c) it is capable of industrial application
d) it is not in an excluded group of ideas

Thus, the invention must never have been made public in any way, anywhere in the world, before the File date. An invention involves an inventive step if, when compared with what is already known, it would not be obvious to someone with a good knowledge and experience of the subject. Industrial application means that it must take the practical form of an apparatus or device, or a new material or substance, or method of operation. And no matter how ingenious, unusual or beneficial an idea might be, it is not patentable if it is

i. a discovery
ii. a scientific theory or mathematical method
iii. an aesthetic creation (music, art or literature)
iv. a method for performing a mental act or game
v. a computer programme
vi. a new animal or plant variety
vii. a treatment of the human body by surgery or therapy.

When an invention is created it belongs to the inventor unless his rights have been given to somebody else, for example by means of an agreement or contract of employment. If an employee makes an invention in the normal course of his duties as an employee then the invention will probable belong to the employer. If the invention resulted from collaboration between two or more people, each person may be entitled to a share in the ownership and should be named in any patent application.

Industry holds by far the largest number of patents, and the pharmaceuticals continue to use Patents to product their products, but others are beginning to question what Patents actually do for them. I think the question arises because when the Patent is granted it is published – competitors then find out what the latest thinking is. Not that they then use it directly, but it helps their own thinking to leap frog along......... thus taking early advantage of someone else's research spending, without indulging in espionage.

The Patent Office will provide a free booklet, the Forms, and guidance on Patents, Design Registration and Trade Marks. Write to

The Patent Office,
Cardiff Road,
Newport,
Gwent NP9 IRH
or phone 0645 500505.

Members of the NAGC Youth Cafe should be able to discuss their ideas on the Internet as if they were not in the public domain, provided they have all agreed to keep the ideas private to themselves, and have appropriate levels of password security. It is therefore in your own interests not to let others see or have those passwords.

CHAPTER 11

HEADWAY

At one time, I was a Volunteer Carer and Trustee with the Headway charity in Gloucester, assisting the attendees with reading, writing and numeracy tasks, and initiating fund raising activities with a self generated presentation. I have thus assisted the brain injured, helping them to prepare lunch with regular cups of tea, play games, and make bus timetables and money easier for them. I worked with an ex bank clerk who, in 1998, could not work out the value of decimal coinage and kept mentally reverting back to the old pre 1971 coinage. He had to trust cashiers in shops to take the right money from his proffered hand, so I wondered whether it might be possible to put the value in bigger letters on today's decimal coins. It might not have helped him, but that was all I could come up with to help him, so I just kept repeating the teaching process. I note the new £1 coin is very difficult to read, especially under overhead lighting because they are so reflective. These coins are also similar in size and reflection to the 10p coin and I wonder how blind people manage.

Learning how to educate others, was interesting. One of the other regular attendees was a young man called Steve, who had been in a major car accident. He had been an active young man up to that point, but was severely brain injured after that. Medicine can do so much these days that people like him can be prevented from dying. In the older days he would have died, and thus not lived to know his fate.

He was permanently in a wheel chair with a table in front of him. He struggled to eat his meal and often failed to put a fork into his mouth properly, jabbing it into his lips. I made him a support for his table to prevent the plate sliding around, and the senior doctor, Joe Willison, in charge of the charity suggested I contact Remap and join them to help others too, which I did in 1999.

But more than that, Steve was unable to hold his penis to go to the lavatory, and had to ask one of the female helpers to hold it for him. I did that once for him when I became aware of his plight – it brings it home to you about how much carers have to do. He also suffered from a head jerk which prevented him from easily following the TV, and I wondered just what occupied him during the day. He must have felt very lonely and distraught with his condition, to the extent where he kept asking anybody around him to help him commit suicide because his hands and arms were too jerky and he could not do it for himself.

He was about 30 years old, and likely to live for many years yet, in such a disastrous state, no future happiness for him. I asked the local MP once to support the Assisted Dying Bill, because of Steve's plight after I described it to him, but he refused to consider it. He also said that he was concerned for the vulnerable, but Oregon has an assisted dying bill and they do not have so many wanting suicide. So who is vulnerable? Do MP's and Lords think that there are so many evil people around that one's children or grandchildren are going to insist they die, just to get their money? Where-ever do MP's and Lords get such horrible thinking? I was very angry with my Labour MP and I challenged him to go to Headway and see Steve and tell him "hard luck chum", but I never heard back from him after that.

I later asked Lord Jaffe to do something about it, but he thought the best way forward was to just go for those who were terminally ill and likely to die within about 6 months. Even that simple Bill was dismissed! Not enough in my opinion – Steve's life and long term happiness matter and he should be allowed to die gracefully. Do remember that no amount of palliative care will help Steve, for he is not in great physical pain, but he is in a psychological hell.

What do his family think about the Government's lack of interest here? I can imagine them feeling very anti Government for Steve was committed to them for years – a potentially long period of difficult caring, during which they would have been forced to endure his continuing demand for suicide. Perhaps in the end he gave up, and became severely mentally ill, or very depressed – yet another problem the Government did not think about - "hard luck chum" they would presumably say when they visited him – if they had the guts to so expose themselves! To the family too.

CHAPTER 12

HUMANISM

I am a Humanist and I have an approach to life based on moral humanity, real data, and reason, and I do not want nor need a God or any other type of superstition to understand the world or universe around me. I recognise that moral values are based upon human nature and experience alone. Real data is collecting factual evidence about the world whether animal, human, vegetable, or other mineral resources. Humanists believe in evolution, that it is a fact based upon real data, and appreciate how wonderfully environment has trained it, and us. We want a world where everyone lives cooperatively on the basis of shared human values, respect for human rights, and concern for future generations. We campaign for a secular state, and challenge religious privilege. A Humanist loves all, trusts a few, and does wrong to none.

I believe in individual rights and freedoms, but that individual responsibility, social cooperation and mutual respect are also important. I believe that people can and will continue to find solutions to the world's problems so that quality of life can be improved for everyone. I am positive, gaining inspiration from our lives, art and culture, and a rich natural world.

The word humanist also means someone who:

- trusts to the scientific method when it comes to understanding how the universe works and rejects the idea of the supernatural (and is therefore an atheist or agnostic)
- makes their ethical decisions based on reason, empathy, and a concern for human beings and other sentient animals
- thinks that, in the absence of an afterlife and any discernible purpose to the universe, human beings can act to give their own lives meaning by seeking happiness in this life and helping others to do the same.

See also
https://humanism.org.uk/humanism/how-humanist-are-you/

Current UK law and government guidance discriminates in favour of religion in requiring daily acts of 'Collective Worship' in schools, and in favour of Christianity in requiring that for schools without a religious character, the majority of these acts of Collective Worship should be 'of a broadly Christian character'. Sixth form pupils and parents on behalf of younger pupils have an absolute right to withdraw from compulsory Collective Worship.

The UK is the only country in the world to impose Christian worship in state schools. In demanding Collective Worship in schools which will typically have pupils from a wide variety of religious backgrounds and none, the law is incoherent: a school can do many things collectively but, lacking a shared religious faith, it cannot worship collectively. It also ignores the right of children, below the sixth form in England and Wales and of all ages elsewhere, to freedom of belief and conscience by only giving to parents the right to have a child excused from worship or to withdraw the child from school for an alternative form of worship, for Government still insists that such worship must continue.

See https://humanism.org.uk/campaigns/schools-and-education/faith-schools/

Parents have an explicit right in the European Convention of Human Rights to bring up their children in the religion or belief of their choice without interference from the state. However, they do not have a right to state funding for confessional religious teaching or 'faith' schools that are in line with their own beliefs.

Humanists think that state schools should not be allowed to choose pupils on the basis of religion, discriminating in access to a public service that should be open to all. We think that state schools should not be free to select teachers and other staff, or to select governors, according to their religion. Humanists are concerned that the proliferation of state-funded religious schools is making for a more segregated future, particularly as religions whose believers tend to come from particular ethnic groups gain more state-funded schools. When studies show that religious selection for pupils also results in socio-economic selection, we think the social case against religious schools is even stronger.

Humanists are committed to secularism – the principle that, in a plural, open society where people follow many different religious and non-religious ways of life, the communal institutions that we share (and together pay for) should provide a neutral public space where we can all meet on equal terms. State secularism (where state institutions are separate from religious institutions or the state is otherwise neutral on matters of religion or belief) guarantees the maximum freedom for all, including religious believers. In such a state, no one should be privileged nor disadvantaged on grounds of their religious or non-religious beliefs. In all schools Humanism should be taught, for it is a good way of life.

The UK and the totalitarian theocracy Islamic Republic of Iran remain the world's only two sovereign countries to give automatic seats to religious leaders in their parliaments - automatic Church of England Bishops in the UK House of Lords.

Why should you go to heaven merely because you believe in a God, or go to hell because you do not, or be the subject of a death threat because you want to leave? As a Humanist, I think there is no God, nor heaven nor hell – immortality means only memories of a person to me, either of a surviving individual or in a book or on film, and not at all a heaven or a hell.

Surely of far more importance is the need to be truthful – it should be remembered that Christianity supported slavery in the UK for over a thousand years before a humanist outlook in the UK adopted the Slavery Abolition Act in 1833. Further, over 50% of the people in the UK no longer believe in a God, and at last Humanists are now allowed to be present at the Annual Cenotaph remembrance in November.

Jesus Christ was a fantastic, independent and courageous man who changed the World with His thinking - I truly respect Him for being so wonderful, but I just think He was not God in the sense that He oversees every individual soul in the World. I am sure that Jesus Christ would take pleasure in our individual freedom to think for ourselves – just as He did. And so, as a Humanist, I try.

I am intelligent enough to think and understand that there are not just 3D (three dimensions) in space, but that rotation around those three axes gives another three dimensions adding up to 6D, while if you add time as well you get 7D in total as can be seen in many robotic machines. I also understand complex numbers, integral and differential calculus, electrical phase angle, high integrity systems and software, Einstein's special relativity, some knowledge of artificial intelligence, and so on – yet I am considered irrational (or even evil according to one Archbishop of Canterbury) when I challenge the idea of a God. And

I think the Devil is – King Henry 8th obviously thought that it was a greater sin to be divorced than the crime of murdering a wife!

Illegal madrasas and yeshivas have been operating in English cities for decades, with many styling themselves as supplementary schools. Both local and central government have been fully aware of the existence of full time illegal schools, but until recently, neither has shown a willingness to act out of fear of offending religious communities. However, HumanistsUK work with whistle-blowers and journalists, brought to light a huge national scandal, as it showed how many children were being denied a future, or even an education in basic subjects like English and Maths, in favour of all-day religious education from untrained teachers in unsafe buildings. This work encouraged Ofsted to set up an illegal schools unit and prompted The Prime Minister to promise a 'crackdown' on illegal schools where extremism and fundamentalism were flourishing.

The house arrest of Galilei Galileo for believing that the Earth went around the Sun was a falsehood by the Pope of that time! Have you ever worked out how fast the Sun would have had to go around the Earth if that were the real case – an enormous 1.2 million mph (1.9 million kph) given that the distance to the Sun was estimated at only 4.5 million miles (7.2 million km) in those days. The Pope jailed Galileo for 10 years!

Confucius said "We have two lives and the second begins when we realise we only have one!"

"We only live once. This is not a drill. There is no second life to come. Even the future in this one life is unpredictable and there are no guarantees about what will happen next. So it makes sense to enjoy life now." From Alice Roberts and Andrew Copson in "The Little Book of Humanism", ISBN 978-0-349-42546-7.

Charles Darwin wrote: "As man advances in civilisation, and small tribes are united into larger communities, the simplest reason would tell each individual that he ought to extend his social instincts and sympathies to all the members of the same nation, though personally unknown to him. This point once reached, there is only an artificial barrier to prevent his sympathies extending to the men of all nations and races." Humanists do not discriminate to any human being, for the genetic code is the same in all of us.

Bertrand Russell wrote: "Love is wise. Hatred is foolish. In this world, which is getting more and more closely interconnected, we have to learn to tolerate each other. We have to learn to put up with the fact that some people say things we do not like. We can only live together in that way. But if we are to live together, and not die together, we must learn a kind of charity and a kind of tolerance which is absolutely vital to the continuation of life on this planet."

"A humanist view of society is based firmly on the benefits of free thought, free expression, and the free exchange of ideas. The principle of freedom of thought and expression should be extended to children too. No one – no matter how young – should ever be forced into following a religion or other tradition. Children must be taught HOW to think, not what to think." And "Many religions contain warnings about curiosity: do not ask too many questions, do not ask for evidence, do not allow yourself to doubt. But for humanists, curiosity is a virtue." Andrew Copson and Alice Roberts.

CHAPTER 13

REMAP

As a volunteer Engineer I enjoy working for free for the charity called Remap Gloucestershire (www.remapglos.org.uk). All of us Remap engineers collaborate with OT (Occupational Therapists) and Physiotherapists, creatively producing custom one-off equipment for the disabled, to improve their independence and quality of life, without cost to them. This helps their carers too, and we are often the last resort in helping somebody, for we start where all the manufacturers brochures finish. We are carers too.

97 of my own designs have been manufactured so far with some help on some of them from my RemapGlos colleague, Ron Crumpler, particularly for manufacture engineering and discussion help from the same wavelength. Another 51 cases (making a total of 148 cases) proved to be better and quicker as advice only – in other words I proposed another client focussed solution which was quicker, most useful, and was accepted, which only required me to think about a solution, but not manufacture it. For example, a mentally disabled client kept pulling tissue of the toilet roll until the roll was empty and then he blocked the lavatory with it, so the solution was to suggest a single tissue dispenser, which was available to order on the internet and which the care home bought.

Remap does not charge the client for any of this work, and we are supported by Gloucestershire County Council with a £1000 annual grant to cover a third of our expenses at about £3500 a year. We also receive financial help from others, like The Barnwood Trust, The RAOB Brockworth, Zurich Insurance, Charles Irving Trust, Cotswold Fosseway Lions Club, Cheltenham Masonic Lodge, and many clients who wish to donate. We also receive support in kind from several businesses around Gloucestershire including Grail Engineering, Emberson Group Ltd, Partington Engineering Ltd, Cotswold Foam Gloucester, GIS Healthcare, Cotsweld, Exception PCB Tewkesbury, Plastim Ltd Staverton, Nick's Timber, AJ Lowther and Son Ltd, Wheatley Printers Ltd, Midwest Mobility, National Star College, Gloucestershire College, and Hucclecote Rugby Club where we hold our meetings. I was Publicity Officer for Remap Gloucestershire for several years, and collected many jobs as a result of my presentations to many charities including those supporting MND and MS (Motor Neuron Disease and Multiple Sclerosis).

Examples from my files include:-

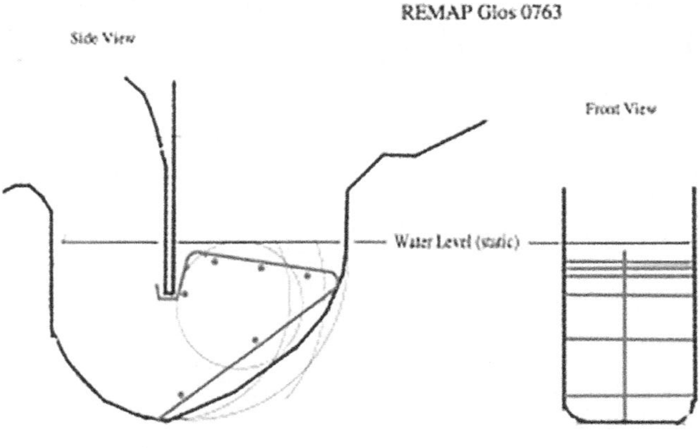

The above Remap solution of mine, was required to prevent an Alzheimer's victim from dropping her waste plastic nappy into the lavatory, and blocking the pipes well downstream. The drain clear out was very expensive, so I saved the client a load of money as a result of my design. I had to put my hand into the lavatory to find out the wall thicknesses and thus to design the underwater grid as shown. My finished article is shown on the window ledge before insertion into the lavatory.

The above Remap case was for an elderly gentleman who had had a knee operation which locked his knees into a straight (standing) position, and he could not get down onto the lavatory as a result. The hospital advised what height and slope was required of an Elsan, and I designed the frame and constructed it with Ron, a Remap engineer colleague who was good at thinking about quality manufacturing. The client was able to happily use the new lavatory for himself, without the aid of a carer – it had two toilet roll holders as shown.

The elderly woman could not get down to the dog bowl on the floor, and had a carer come in twice a day to feed the dog for her. She asked Remap to solve the problem, and I came up with the above extended handle centrally located over the bowl so that it would stand properly with food in it. She was able to lift the bowl off the floor and onto her worktop to load it wth food. She had to be able to reach up to the handle, so I designed it not too tall. The carer no longer had to come at all, so I saved Gloucester County Council some effort, and the elderly woman was more independent and pleased. Remap Engineers work for this sort of thank you result.

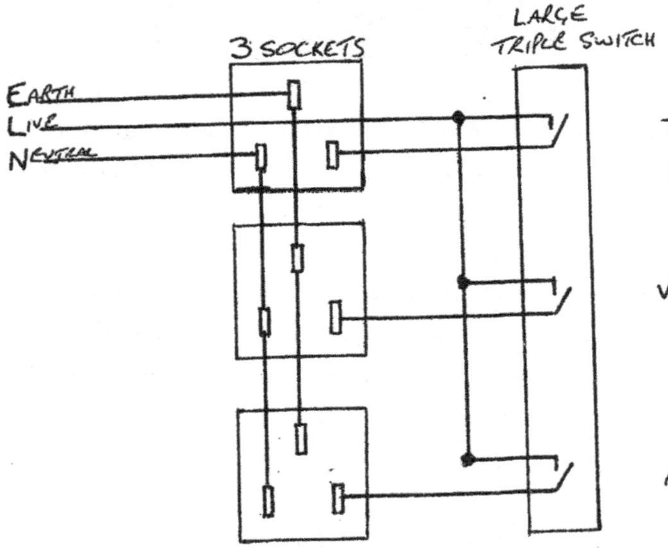

The above job was for an elderly woman in a care home who needed a Carer to come and change from TV to Video or Radio, and back again, for her. She could not switch on or switch off these things herself, so I made a big triple switch with a long cable from the wall electrical socket, that she could use herself. She only wanted to watch BBC1 on the TV so switching on and off was all she wanted to do for that. The need for a Carer to help her was much reduced, and the woman was last seen gaily switching her utilities on and off in rapid succession!

The above job was for a teenager who wanted to ride her bicycle with her friends around the village. She had a right arm which was 150 mm (6 inches) shorter than her left and her hand was also deformed with short fingers. So I modified her bike to put a dual brake lever on the left with cables to front and rear, and balanced the front and rear brakes with an adjuster, both of which I bought off the internet for the local bike shop did not have these. An extension to the right handlebar was made by Ron Crumpler, and fitted by me, which she was able to reach to control her steering effectively.

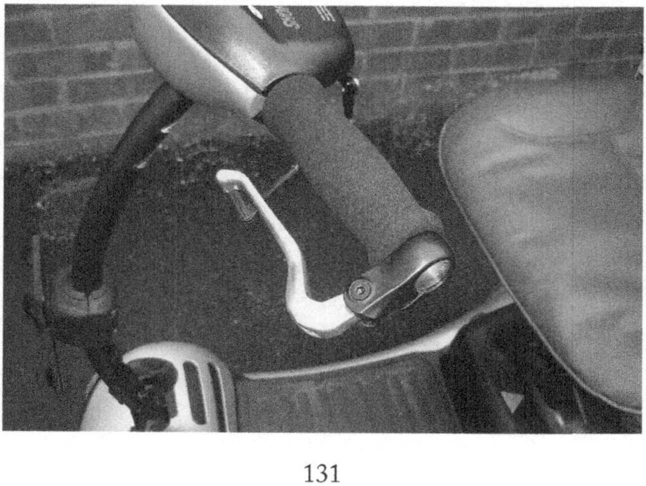

The above job was for an elderly woman who used an electric scooter to get around. She had used her right and left thumbs to control forward and backward motion respectively, but was no longer able to do so because of arthritis. However, she was able to use her fingers to control her speed using bicycle brake levers which I then fitted to the vehicle. A linkage was made which translated the brake lever movement into thumb lever movement, and to ensure her right hand still controlled forward motion, and left lever for reverse, so the electrical connections under the head of the scooter were swapped over.

The above job was for an elderly woman who was finding it difficult to remove the ashes from her coal fire. I designed a frame which would clip into the ashes tray which she could bend down far enough to lift. The frame could not be made taller, because otherwise she would not have been able to lift the ashes tray high enough to tip them into her dustbin. This case was featured in the local Citizen newspaper too.

The above job was for a very short 91 year old woman who was recovering from a fall, and needed help to walk around her home. She had a trolley for which I designed and made a gutter frame to support her arms, based upon the handlebars from a time trial bike which had low height forearm supports. The reason I used the time trial bike idea (suggested by my daughter), was because the arm supports were very low above the top tray of her trolley, compared with the usual NHS ones which were over 150 mm (6 inches) too tall.

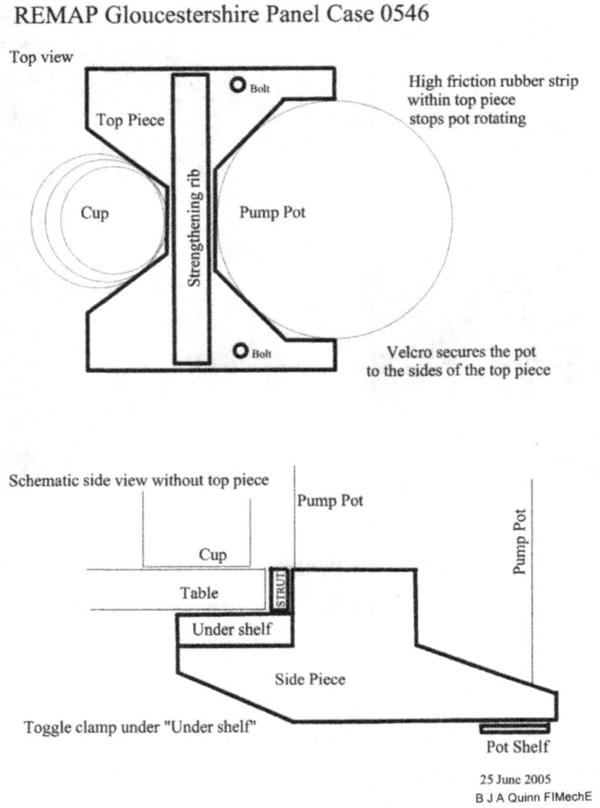

This case was for a young woman who wanted to be able to help her visitors with a cup of tea, without needing her Carers to pour it for her. I designed it and Ron Crumpler made it – Ron's idea was to mount the pump pot halfway down on the table so that splashing while pouring was minimised – the pot spout was thus only a little higher than the top of the cup. It worked well, the Carers now only having to fill the pot with tea and then give it to her.

This case was for a short woman who could not open or close her window, because she could not reach the top lever, so I measured and drew a design which Ron and I made. This connected the two levers as shown. She was happy.

This case was for an elderly woman who could not hold her cup of tea and at the same time use her walker to go from kitchen to sitting room, so I fitted a bicycle mount on her walker as shown. It made a lot of difference to her she said, and thankyou. These thankyou's are always precious to me.

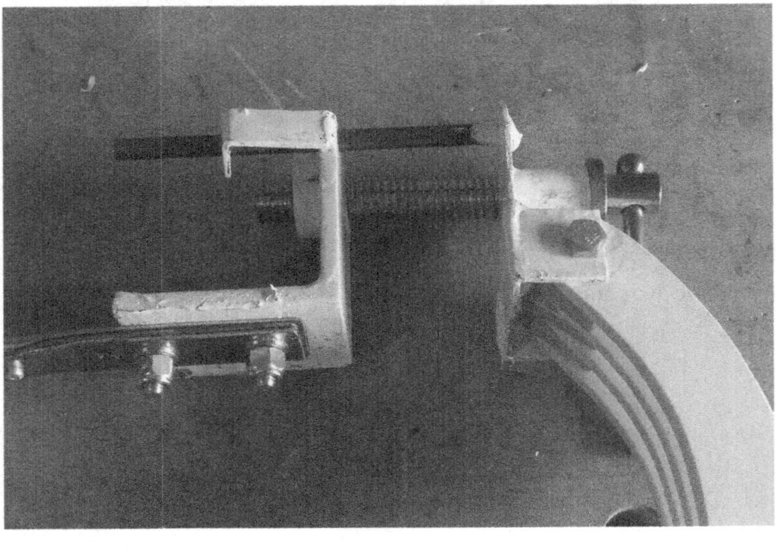

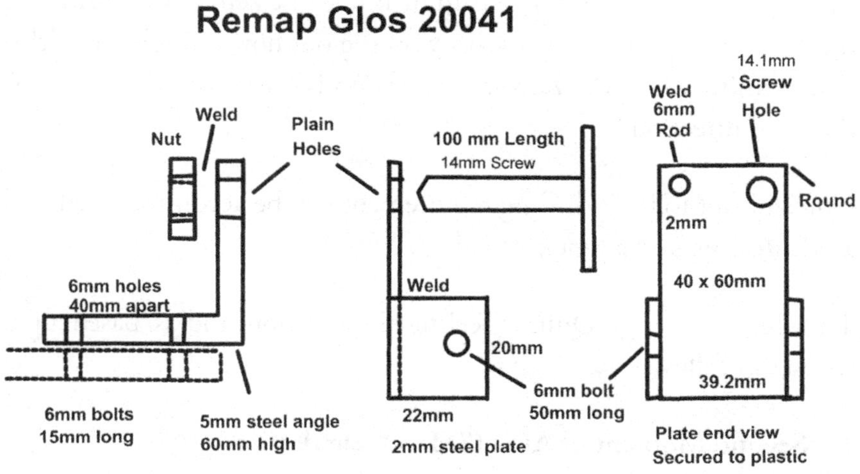

This design was for a 63 year old who had Motor Neuron Disease and she could not lift her head off her chest to swallow properly without a frame made for her, but the adjustment was too coarse and therefore difficult to set up for her to be comfortable. My design was made to finely tune (by use of a G-clamp screw thread) her head position so that she could drink with her chin off her chest and yet not too high. I had to make a further modification relative to the drawing to improve its stability as shown in the photo. It was easy for her carers to adjust, and she said it was "marvellous"!

Wheelchair Frame Drilling Rule

This Rule is only applicable to wheelchairs or walkers out of warranty, since warranty is cancelled by a manufacturer or retailer for any drilling, no matter how little stress is caused.

RULE: Use a 6 mm drill normally into 22mm tube, and drill only from the side of the wheelchair or walker, not from front to back because that is the high stress hole position.

The logic of placing any hole is to drill it at the zero stress position in the tube. This particularly means working out how it might bend thus to determine where the zero stress position is – that is, sideways across the bend direction.

Note: a front to back drilling requirement can be accommodated with a side drilling and a bracket made to suit.

This Rule is a Jim Quinn Technical Direction and is based upon stresses as follows:

1) **Second Moment of Area (SMA) Calculation**
   22mm OD tube, 1.25mm wall thickness

   a) Cross Sectional Area (CSA) of tube = 81.5 mm^2

      CSA of tube with two 6mm holes = 66.5 mm^2 (-18%)

   b) Second Moment of Area (SMA) = Σ (y^2.dA) calculated by taking slivers of a shape, and is needed for bending or buckling stresses.

      SMA for circle = (π.D^4 / 64) about a line xx through the centre of the circle. D is the diameter

      The centroid, G, is the centre of the circle.

      SMA for rectangle = (B.D^3/12) about a line xx through mid height. B is rectangle breadth, D is height. SMA for remote rectangle = (B.D^3/12 + CSA.h^2) where h is distance from xx.

   c) Tube without Holes

      Second Moment of Area of 22 mm tube 1.25 mm thick = 11499 - 7098 = 4401 mm^4

Note: SMA of 20 mm tube = 7854 – 4604 = 3250 mm^4 (-26%), and thus why 22 mm is normally used for bending and buckling reasons.

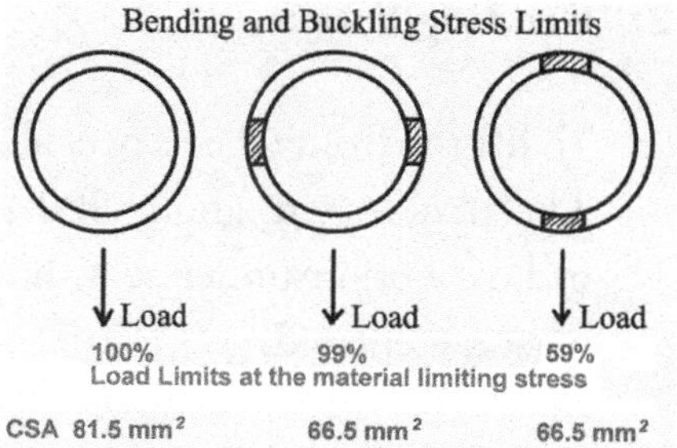

22 mm OD tube with several 6 mm holes in a line

d) Tube with Holes centred at Zero Stress Position

SMA of two 6mm holes at zero stress position = $2 \times 1.25 \times 6^3 / 12 = 45$ mm^4

SMA of zero stress holed 22mm tube = 4401 - 45 = 4356 (99%)

Note: hole 0.5 mm off-centre is only 0.04% SMA reduction, to 4354 mm^4 = 4354×10^{-12} m^4.

e) Tube with Holes at Maximum Stress Position

Assume holes are rectangles in the tube cross section.

SMA of two 6mm holes at max position = $2 \times 6 \times 1.25^3 / 12 + 2 \times 11^2 \times 1.25 \times 6 = 1817$$mm^4$

SMA of max stress holed 22mm tube = 4401 - 1817 = 2584 (59%) mm^4

Note: hole 0.5 mm off-centre is zero SMA reduction, so SMA = 2584×10^{-12} m^4

f) So, logically, drill holes at zero stress position.

2) Bending Forces

To lift the front wheels over a kerb, the handles are used in bending. The handles are about 800 mm high, length is about 500 mm to box section under seat, but about 300 mm from hand grip to person Centroid.

Probable Handle Force = 111 N each (50 lbf total).

So, Maximum Bending Moment = 111 x 300 / 1000 = 33.3 Nm

Max Stress = M.y / SMA = 33.3 x 0.011 / (4354 x 10^{-12}) = 84 x 10^{6} Pa = 84 MPa with 6mm holes at zero stress position or = 142 MPa with 6mm holes at max stress position. So, logically, drill holes at zero stress position.

84 MPa is safely 43.5% of the 193 Mpa material limit, and is with 6mm holes centred on zero stress line with 0.5 mm (20 thou) accuracy.

3) Tube in Compression

2 rear wheelchair vertical struts share compressive load of, say, 20 stone (= 650 N each strut).

So, compressive stress = 650 x 10^{-6} / 66.5 x 10^{-6}= 9.8 MPa with 6mm holes.

This is only 5.1% of stress limit – they could take 20 times more – 400 stone! So that 18% reduction in CSA due to the 6mm holes, is no problem at all in compression.

4) Buckling Load

Euler formula = π^2 x E x SMA / L^2

Seat 350 mm high, so strut lengths to small wheel centre are a maximum 250 mm long, so L = 250 mm (worst case)

Min SMA = 2784 x 10^{-12} m^4

Mild Steel Yield Point is 215 GPa & Crushing Stress = 300 MPa. Note: CSA with holes = 66.5 mm^2

Mild Steel Young's Modulus, E, is 200 GPa = 200 x 10^9 Pa

So Euler buckling load = 3.1459^2 x 200 x 10^9 x 2784 x 10^{-12} / 0.250^2 = 87926 N = 87.93 kN

The Rankine Gordon correction gives an actual Buckling Load of 1 / [1 / 87926 + 1 / (300 x 10^6 x 66.5 x 10^{-6})] = 16261 N (16.26 kN) per strut

Two rear struts could thus support 32.5 kN (1000 stone), so 20 stone applies only 2% of allowable load.

Repeat: The logic of placing any hole is to drill it at the zero stress position in the tube.

CHAPTER 14

SOFTWARE

I am computer literate with much experience of internet and email, Windows, LibreOffice, PowerPoint and presentations, Z80 Assembler, Scheme P, Fortran, Visual Basic, C++, 2D Total CAD, the Veribest Electronic Chips Database, and Simulink (Functional Control Systems) Computer Aided Design Engineering package.

QUALITY

There is no such thing as software reliability, but there is software Quality. Reliability is in the computer hardware (the electronic chips and the PCB's) perhaps due to vibration or over temperature, but not in the software itself for that is not going to fade away nor change through use or time. EMI (ElectroMagnetic Interference) can affect the software recorded in memory, but it is the memory chip that is susceptible not the software. However software Quality is in the construction of the software, and loops particularly are prone to interrupts and step out GOTO's. These can cause havoc, which is why static code analysis software like MALPAS and SPADE (developed by RSRE Malvern – Royal Signals and Radar Establishment) that can analyse the target software to illustrate whether the Quality was good enough or not, is so essential.

This was particularly relevant to the Mull of Kintyre crash on 2nd June 1994 which killed 29 people in an RAF Chinook, for it had been previously declared to ModUK that they should not buy that FADEC software from Textron (Boeing's subcontractor) because it was so poor in Quality! The RAF Pilots were initially blamed, but that was so very wrong, and they were finally cleared in 2011 – it was the engine software that was at fault.

There is a strong economic case for identifying defects as early as possible, as the cost of correction is very significantly increased the later the defect is discovered, but the process is more than just the Coding – it involves the Quality of the Customer Requirements Definition in local language (English for me) and the Design Team Specification and Functional Design plus the closing Testing Procedures before Coding even starts. Note that Requirements do not have to be complete before Design and Coding starts because it could be done in partial releases, and there may even be a prototype phase during which Requirements and/or Design are evolved.

Formal Methods like VDM and Z have been developed to help improve the quality of the Specification. The main advantage of Formal Methods is that the mathematical Specification produced can be proven to be correct giving more confidence in the integrity of the result, but the Requirements can only be written in local language like English, and the Customer is unlikely to understand VDM or Z, so the Specification written that way does not help the Customer to understand what the contractor is going to do. In such circumstances maybe prototype demonstrators will be required. However, the high costs of introducing Formal Methods Specifications has deterred Software Engineering companies, but they are more useful in Integrated Circuit design, for IC's are used in many different companies products and a single defect (like the Intel Pentium calculator suffered) can thence be spread world wide. It might help the writers of the Requirements in normal colloquial English, if the Formal Method Specification was reverse engineered from its mathematical ladder logic into readable English for those who cannot read the Formal Z or VDM. I propose that software be developed to do this automatically, so that Requirements can be logically portrayed.

Ada has been specified by MoDUK for use in Defence Equipment for Quality and Readability reasons, but even that has to be written in Safe Ada, which is a safety critical subset of Ada.

The evolving capability of the modern computer and its software is shown in this next chart. The aerodynamic problems listed show how the software has evolved, for the more capable computer allows its complexity and quality of solution. The Cray computers were supercomputers with staggering ability compared with my early days on the KDF9, and it was able to compute over very large and detailed geometric grids compared with the very small and slow simple inviscid flow computational fluid dynamics, CFD, for the BS143 and RB199:

DEVELOPMENT OF AERODYNAMICS AND COMPUTERS

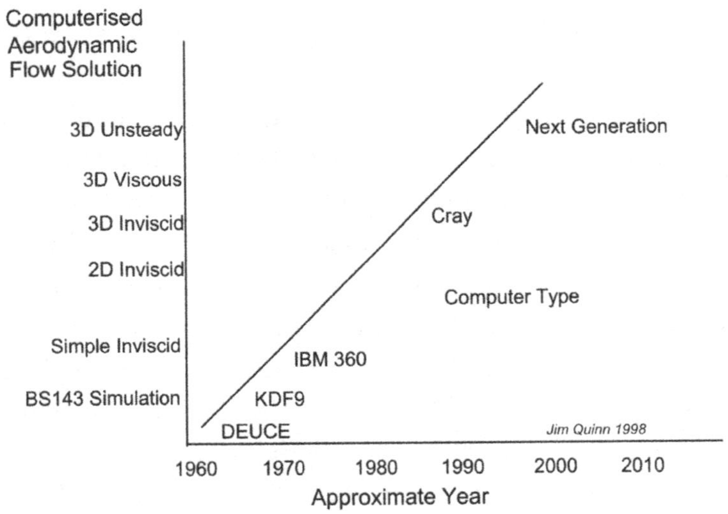

SCHEME P AND FORTRAN

Scheme P was a language made specially in Bristol Siddeley for Performance Engineers to compute the pressures and temperatures throughout a gas turbine engine on the KDF9 computer – such simulations had not been possible on the DEUCE computer. To ensure that calculations were consistent across all employees, the language was based upon the idea of "bricks" as we used to call them – thus there was an intake brick, a compressor brick, a combustion brick, a turbine brick and a nozzle brick. All that then had to be done was to link all of the bricks in the correct order and with the correct transposition of data to produce the simulation. Necessarily, this was more complicated because I had to add in the various airflows leaking or deliberately flowing from the compressors to the turbines on both BS143 and RB199 engines. I also had to add in limitations on HPCompressor delivery temperature so that we could compute control functions. Then feed the simulation with the Mach number and altitude, the spool speeds and mass flows at each part of the engine, and the pressure ratios of the compressors to compute the rest by assigning a combustion temperature. Bristol Siddeley under Neville Quinn, the Chief Performance Engineer, and not a family relation to me, had defined the need for Scheme P for when the KDF9 computer was first obtained in about 1965, and all I had to do was link the bricks in the right order and with the right data transpositions to produce the steady state simulation. But those bricks had to be written first, and they were well defined by Arthur Brown and Dickie Burden based upon their learning, like mine, of thermodynamics and fuel/gas chemistry. The accelerations and decelerations were simulated by me, by introducing the component mechanical inertia into the turbine power requirements, and I then ran that simulation to produce the IPCompressor chart shown in an earlier chapter.

One of the problems I had when creating the performance inputs for the 1969 RB199 specification were that the other users of the KDF9 had to be stopped while I ran our work because our Scheme P only had Brick

28 to optimise the off design work by matching turbine and nozzle capacities. This resulted in pressure to introduce a matrix solution procedure which was eventually done in Brick 41. The other problem that Brick 28 gave was that it converged badly – some oscillations sent the graph read search off the graph, so I had to widen the graphs well outside their normal operating ranges. One of the important things in software is to make sure that it works!

To allow the Airframers to use the simulation, my Scheme P simulation was converted into Fortran in 1971, for Scheme P was not easily transferable to others computers. We did not have Fortran right from the start because the KDF9 did not have it available, but Bristol bought an IBM360 in 1971 and Fortran had been created by IBM in the 1950's. The simulation was performed using Fortran subroutines instead of KDF9 Assembler bricks. Hilary Castle did some of this conversion with me. Fortran was also used for the FAP (Full Analysis Programme) that computed the engine efficiencies from the engine test measurements and which I helped Bill McDougall to create for Brian Yates' team to use, based upon my knowledge of simulation software.

BASIC AND Z80 ASSEMBLER

In 1985 I bought a Sinclair Spectrum which used the software language Basic and had a Z80 CPU. I learnt Basic from it and used it to programme Car Rally results, but I found it was too slow sorting cars into the right results order, so I then learnt Z80 Assembler and programmed that to Bubble sort them. I had to adapt the routine to move the car number as well as the car time into the right order, for the documented bubble sort procedure only sorts the times. I used it to compute Stage Rally results successfully for a few years, but in 1988 I tried to programme the software for a Marathon Road Rally and failed, for I found it much more difficult to compute road penalties for that Event. This sort of failure leads to evolutionary learning as to how to do it next time.

JQ Rally Result Bubblesort, but I actually wrote it in Z80 Assembler

My Bubblesort algorithm was ***evolved from that published into the italic extra*** because otherwise you would not know which car was fastest to go with the fastest time for example – or 2nd or ….. or 29th etc.

I have shown it here in C++, where n is the number of cars, tim[] is the car times in seconds array, and car[] is the car number array:-

```
void BubbleSort ( int tim[ ], int car[ ], int n )
{ bool swapped = true ; int j = 0 ; int tmp ;
  while (swapped)   {        swapped = false ;
                              j++ ; for (int i = 0 ; i < n – j ; i++)
{ if(tim[i] > tim[i + 1]) { tmp = tim[i]; tim[i] = tim[i + 1]; tim[i + 1] = tmp;
                    tmp = car[i]; car[i] = car[i + 1]; car[i + 1] = tmp;
                                                          swapped = true;}}
                    }}
```

LISP

Lisp was the original language of artificial intelligence which is what attracted me to it. I had a Sinclair Spectrum 48k when I first bought a Lisp compiler, and I did that because many people were playing with the Spectrum to control a mouse that had to walk around a maze, to find the exit. I never managed that myself for I was still learning Lisp, but I did modify the compiler to use apostrophes instead of having to type QUOTE all the time, so I learnt about the way Lisp stores its numbers and sentences. I also used it to solve the famous Tower-of-Hanoi problem - see ISBN 0-201-08329-9 “Lisp”, from which I started learning Lisp.

This involves moving a pile of discs from peg A to peg B using a spare peg C. The discs can only be moved one at a time, and no disc can ever be placed on top of a smaller one. The Lisp programme is quite short so I programmed it into my Lisp compiler, and it is called "tower-of-hanoi N", where N is the number of discs, and you call it as shown:

>(tower-of-hanoi 3) *produces the following output verbatim*
((move disc from A to B)(move disc from A to C)(move disc from B to C)
(move disc from A to B)(move disc from C to A)(move disc from C to B)
(move disc from A to B))
>

and to prove that I have actually run the programme myself, here is the answer for 5 discs of differing sizes, but instead of typing (move disc from A to B) I have simply typed AB:

>(tower-of-hanoi 5) *produces the following summary*
(AB AC BC, AB CA CB, AB AC BC, BA CA BC, AB AC BC, AB CA CB, AB CA BC, BA CA CB, AB AC BC, AB CA CB, AB)
>

The Lisp book also taught me about artificial intelligence in ways to approach a problem – depth first and breadth first searches of a tree as shown are ways to do a job. Depth first means to progress down a tree to find a solution (A B C D then back to B for E, then back to A for F G H and so on), while breadth first merely runs from side to side of all the branches (B to F, or C to E to G, but you can only do this if you know them!), never producing solutions until all the branches have been explored. This latter is the most expensive way to solve a problem.

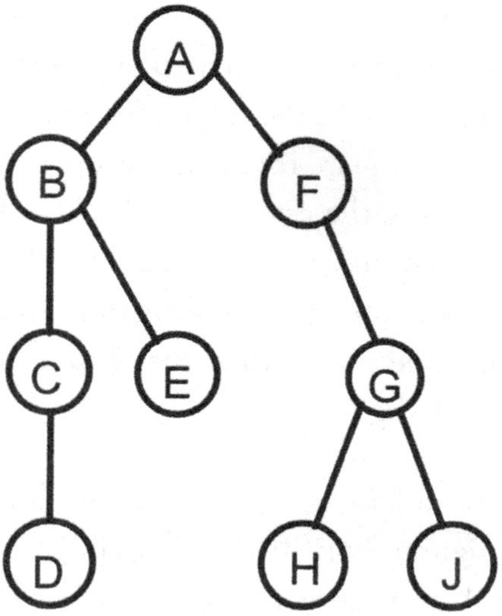

Engineers will normally always perform depth first searches as shown, and will move to another branch if a solution can already not be found at the desired cost or programme timescale. I tend myself to do depth first searches for an answer when asked a question! I retaliate with a breadth first response when I know the sidestep answer well enough – like Parliamentarians do, when ducking the question!

VISUAL BASIC

I write my own software today as well as in years past on Tornado Engine Performance and Behaviour Simulations. Visual Basic is similar to Sinclair Spectrum Basic. An example is my "Long Number Calculator" written in Visual Basic, which calculates to 5010 digits whenever needed. I could make it perform a much higher number of digits, but that has lost its excitement, for it takes too long to do the calculations anyway – it took seven minutes to calculate the 5010 digits of the square root of 3 (and the same time for the square root of 43) on my 1.86 GHz Intel(TM)2 core CPU desktop, with 2GB of RAM.

The calculation procedure that I used was just the same one I learnt when I was at Junior school – put the two numbers so that the decimal points were lined up and add the digits carrying tens as I went. And similarly for divide, but square root became a matter of estimating the answer, then multiplying the estimate by itself to find the difference from that which was required to be rooted. Then halve that difference, add it to the estimate and go round again. This is why, for each estimate multiplying and subtracting, that it takes so long to compute the answer to so many digits.

I found it of interest to look at repeats of digits, and 43 divided by 47 repeats after 46 digits (43/47=0.9148936170212765957446808510638297872340425531 and repeats 9148936 etc), while 431 divided by 471 repeats after 78 digits, and 4311 divided by 4711 repeats after 672 digits – I find this fascinating and you will never see it if you just use the normal 32 digit Microsoft Windows calculator!

Here is my example of the square root of 3 to 5010 digits (and this has no repeats, but 2017 does once in the 16th row, and 2016 four times!):-

1.7320508075688772935274463415058723669428052538103806280
05580697945193301690880003708114618675724857567562614141540
40670302996994509499895247881165551209437364852809323190
23055820679748201010846749232650153123432669033228866506
72254668921837971227047131660367861588019049986537379859
38946765034750657605075661834812960610094760218719032508
31458295239598329977898245082887144638329173472241639845
878553976679580638183536661108431737808943783161020883055-
24901670023520711144288695990956365797087168498072899493
29648428302078640860398873869753758231731783139599298300
783870287705391336956331210370726401924910676823119928837
56411414220167427521023729942708310598984594759876642888
97796147837958390228854852903576033852808064381972344661
05968972287286526415382266469842002119548415527844118128

6534507035191650016689294415480846071277143999762926834629577438361895110127148638746976545982451788550975379013880664961911962222957110555242923723192197738262561631468842032853716682938649611917049738836395495938145757671853373633125910899655424624834787197605235997769192323570220305302840385915414971072429559206706202509520175963185872766359975283663431080150665853710647328538625922260582220510403680270297504798728079461658100417052681940019095733462175943893670249320422691034369812463720111185261084268910299720311202100063507176374582405203847555197279933797614906107894985544223326004018851303631561144886847281589288163245187265066645384877599162576642872111240842068016763517100102943180715515190961642460907039408129216903517492961364004139670431041253632327030922577327960292376597745537095469115742140424230781992327617401906424512454877516862696105333694216213605394604245654140128533007813633449856736406703977342229811961042925534501601405940479547154534548407271737656262366549166640233006013265744070107836858468452313160467754480500402240639911970362218602920238867150711017169400296868759663500040895316214233425227956834067013470185902028360716762147743493449563595808082130442586469468522610908263353008756612603460672195404055984128912975994810000772057440230047673258800091514371489475444879157191294659083570873961515537797640262068370848046072969382719585689759759626104159152657577790782334980567840022901532052138935373775536566427046826874289963441395743666073744445583086477893212985302148197395341478170516614952551763291993699565744522639112519093541386989366817430938226424736926202072990967831154131946484377915459915923928287714695149274026409213645654041644581490201945749419305269002613972646081065071439603206077510594187798284793986195249964165213139715293599421897416647075187235788629466108560170428869605798394052906407430811833388677881562635867156008396760245349229943938

86705975431544294343095725847098821546311126076677406786
457157806064744749975035454455931328654918984933657274762
62974147382356869148378313633612836279038248401638066716
07179848728555842931349226093240565957553651136754644387
83428331346664455418039082189898329462634501617112201692
96194601693206210330397448662316560035667818134431008965
68706742965658177445572685015631594412507361198063472864
662413271936694267129514503975189283134861928413032551884
311690146714237943747283343446524820437960597328706162749
109415968626253025588897684704175020031402622284691048046
98403243790660673774962710251271060206086266628436660501
92890893838320006091228691881836842361693903523126930368
613796811526946997891926635284389927016519825502252923579
36628163456810891307411267837819944977276544829492043880
13017835464992135740327915775077860022964235042759969976
269187384148473947738515635520912209633158390345671894152
943570406530760852901742497681855028852763449495774491104
225024537231560636181035048717591580327917983675100671361
00066217566049573766777112925287302479926720158996418937
762016114118289899682955902511656391422364995062850030696
8129735072517279185983011152775422988518853117635947860031
230637260983698503083108521455957483595557157393566336808
648210787912470510316315695816043487488296148941545853555
619749992968779550992855274235902518103237668616095925525
44319954907420679739157072614245328950428097351037967736
660135917522627061232312589450697634813177754456066603576
22048300690422538949486772952132571976924592767066076581
65582986750565728946588609190296443812850578597594375064
02446062121456085608545249251368328754843438733720201692
31163043270564806013758793685672153508781684407802329206
462316381129177229775272553107516138368213460457779985462
562208196063163601920633096851621623166068346886311662150
132385428739130449512315521636562129518747553097657063995
50298804256473405577851913495330398886430714317850520079

8624246848138297634080217629705645721316219830682936462868222227992381641635317259827321307971419267628691568118183996651510030701082670312866401591831271960403892354068259150459576491836855261480346510928989655996182971640849813570292272962214026375059551994750161810585447401976497159376634238856970198221404541020964290838824886098721887649480083594449193447893133108077774083803 56

and my Computer Monitor Screen Display showing the Square Root of 2, and the About message, is:-

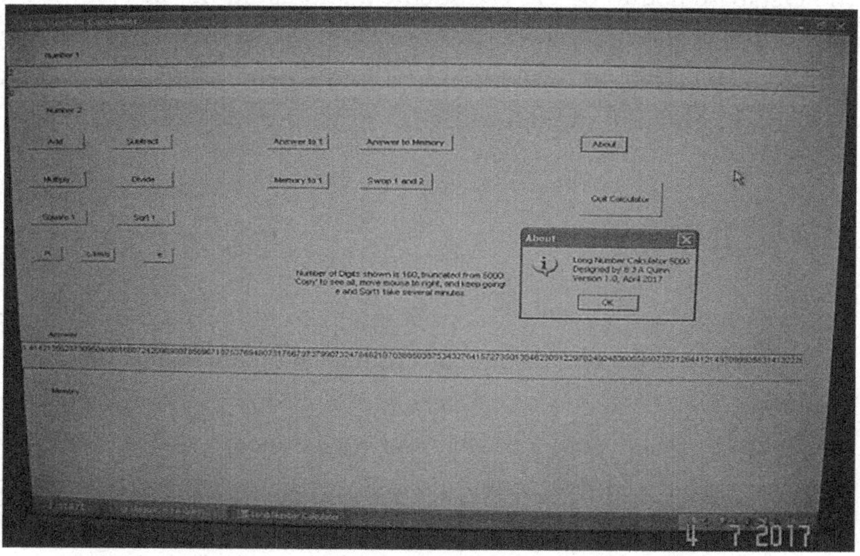

VERIBEST

I worked at Lucas in Cirencester for 3 months, and was asked to investigate the electronic component Database that was called Veribest. It assumed that a component once certified in one electronic unit design could be used in other designs with the same clearance and properties. After my investigation into the varied electronic component usage at Lucas, both in the design office in Cirencester and at the manufacturing facility in Birmingham, it became obvious that the certification obtained in one Project electronic box could not be read across to another Project's box. That was because it had to be certified by test quite separately and may have been a more difficult requirement to meet, let alone that the different placement of that electronic component on its Project PCB (Printed Circuit Board - often glass fibre) might have resulted in a different mechanical vibration environment, thus giving rise to a separate Certification. Therefore that one electronic component had to have, impossibly, two entries in the Database, and so the Database was not good enough, and I said to the Veribest Representative that the Database had to be Project oriented, and could not be assumed to universally cover all Projects with one entry as had been presumed by its designers. This produced an eruption of emails, but I considered that I was right, and did not recommend Veribest for Lucas.

SIMULINK

I experimented with a Simulink package, as would be used in a control system, below. I first heard about Simulink just before I left Rolls Royce, which is why I have said that I needed more time in Controls Design before starting auditing. The photo of my computer screen shows my favoured result and the Simulink logic diagram shown beneath it, shows my chosen values for the PiD (Proportional Integral Differential) controller. The computer screen shows the Simulink diagram on the right and which I have reproduced below more clearly in a printout of that diagram. The left hand side of the screen shows the speed vs time chart that resulted and you will see the slight hunting small up and down movements of speed effect of a good controller, while other values chosen for the PiD controller gave wildly fluctuating speeds. The chart underneath that shows the engine or braking force vs time for the same times as the speed chart – the engine force is positive above zero and the braking force is negative. The external forces (the hill due to gravity, and the much smaller aerodynamic) are shown beneath that.

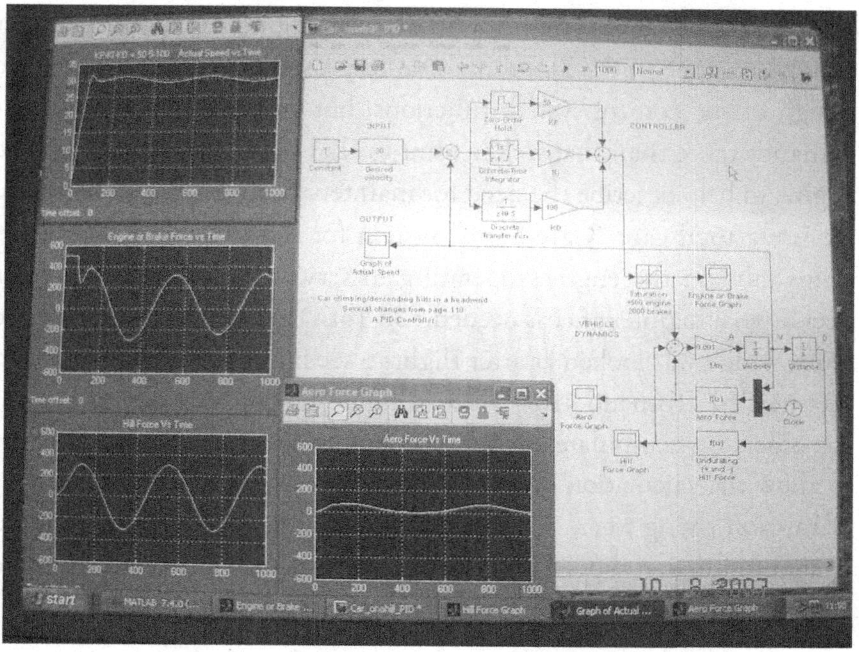

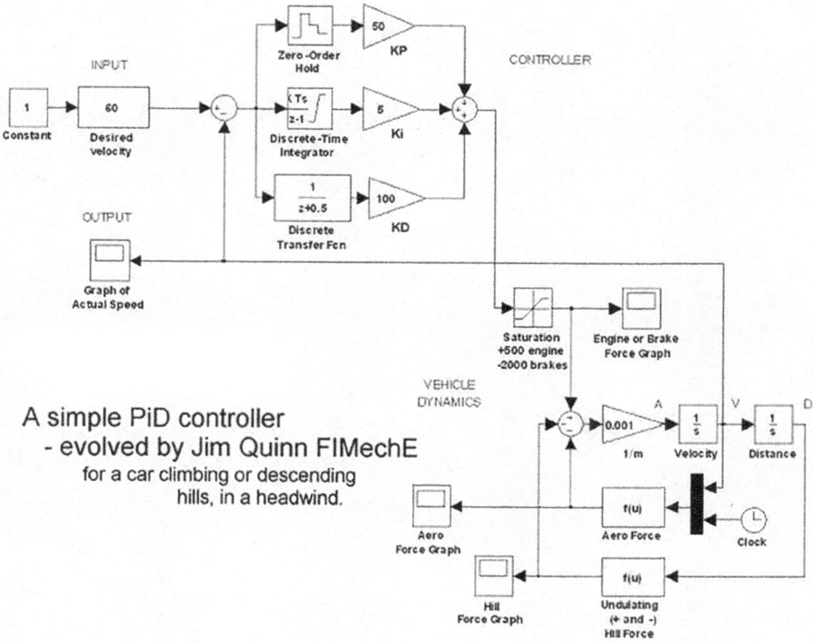

A simple PiD controller
- evolved by Jim Quinn FIMechE
for a car climbing or descending
hills, in a headwind.

However, a FADEC system for an aircraft engine has more than just control to cope with. Simulink is a very useful tool to describe the design of the control system functions, but an aircraft manufacturer demands more than that – they want to read sensors positioned on the engine to help describe the need for maintenance – indeed Rolls Royce plc use a total care "Care Store" system for their civil engines which keeps tabs on the engines operating in service to help the Operator decide what maintenance is needed. On Tornado we had a Maintenance Panel which was looked at after flight to see if any maintenance action was needed, Tornado also had an Engine Usage Monitoring System Recorder to accumulate flight information on a few typical aircraft, to allow the calculation of engine life cycle usage over time. The types of mission flying were also recorded so that life cycle counts could be related to mission type. All of these recordings were collected from the engine control system electronics.

CHAPTER 15

TIME AND TURRET CLOCKS

Ringing in the New Millenium

I rang the bells for several years at St Lawrence in Stroud – and I rang the bells at the New Year 2000 celebrations, and I was in a photo in the "Stroud News and Journal" newspaper as a result.

CLOCKS

Reference Timing of Stroud Parish Clock

I credit My Friend Michael Maltin DFC (earned on Spitfires and Lancasters in the second world war) with introducing me to thinking about the accuracy of a Church Turret Clock as follows.

Given a stop-watch set to a radio atomic clock, what can be learnt about the accuracy of a large turret clock?

The obvious is Strike Time – this is the time of the first Strike of the hour bell, which at Stroud Parish Church is the 20cwt tenor bell and the biggest of the ten bells there. This first strike is the essential time heard "on the hour", and has been heard since the clock was manufactured and installed by Smiths of Clerkenwell in 1878.

But the Strike is not the basic time that the mechanism is keeping, since it is far removed from the pendulum – the recorded twice a day time chart (fig 1 below) shows much scatter, and the clock is shown to be slowing down by up to 5 seconds over the ten day period.

The Pendulum

The Stroud pendulum has a 1.5 second beat (1.5 second to the left, then 1.5 second to the right, to complete a 3 second cycle). The pendulum is temperature compensated, which means that it is made so that the downward expansion of the iron as temperature increases is compensated by an uplifting expansion of zinc.

The Escapement Wheel is a pendulum beat counter – the Escapement is the means by which each swing of the pendulum turns this wheel. The Escapement at Stroud is a "double three legged" type – this was invented by Lord Grimthorpe and fitted by him in 1854 on the Westminster Clock (Big Ben) at the Palace of Westminster, and fitted to Stroud Parish Clock in 1894.

The Escapement Wheel therefore registers the basic time of the pendulum, and by marking a point on the rim of the wheel and timing the markers movement past a pointer, the time (E) can be recorded. The Stroud Escapement Wheel completes a rotation every 90 seconds, and through gears, drives the Main Arbor which contains the Quarter and Hour cams and the clock hands drive. The cams trigger the Quarter (Westminster) Chimes and the Hour Strike, and these thus provide the overall summation of the Escapement Wheel rotations.

The Escapement Time (E)

So, the Escapement Wheel marker point time is recorded on the stopwatch as that which first occurs after the Strike Bell hammer is triggered. This, then, is the basic time of the clock.

The chart (fig 2) shows how this time smoothly varied by 3 seconds in this period of ten days in July 2002. A gradual slowdown occurred, but it was not adjusted during this period in the interests of recording the whole operation of the clock.

Thus the clock is actually slowing in that ten days, by 3 seconds not the 4 seconds (the average of the readings) that the Strike Bell would appear to indicate. Indeed, if readings had not been taken so regularly and for so long, it might have been possible to take an early Strike Bell reading of say 58.9 seconds (see fig 1), and 57.5 seconds nearly 4 days later, indicating that the clock is gaining by 2.5 seconds a week when it was actually slowing by about 2.1 seconds a week.

Recorded Times

There were four timings which could easily be made by stopwatch on the Stroud Clock – these were the H, Q, S and E times, that is the Hour Cam lever release, the Quarter Cam lever release, the Strike Bell Time and the Escapement Wheel time. The H and Q cams are on the same arbor, and the H cam release occurs nearly 30 seconds before the Q cam release. The H cam sets the Strike mechanism to await the Quarter Chime completion, before proceeding with the Strike.

These times can thus be subtracted from each other, in turn, to determine why the Strike data is so scattered.

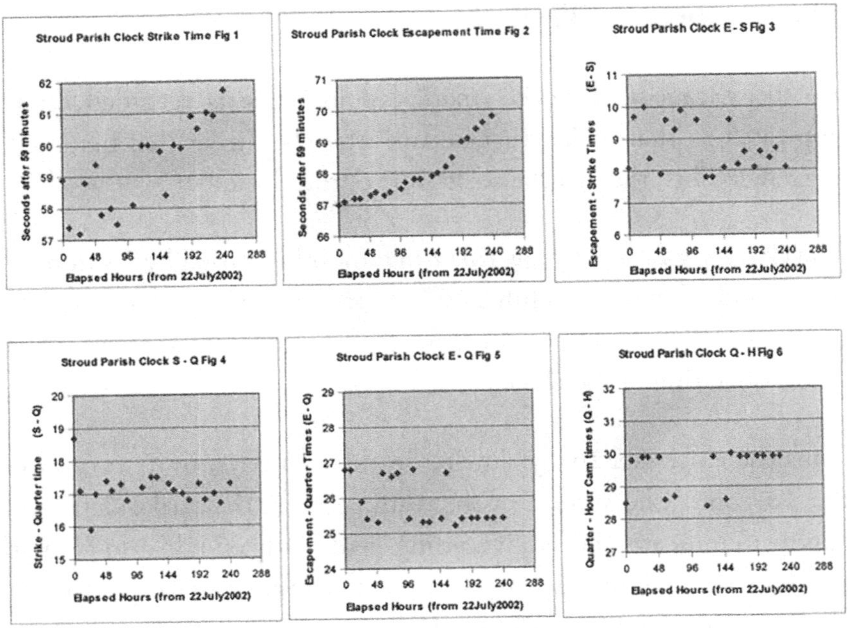

Escapement (E), Hour (H), Quarter (Q), and Strike (S) times and time differences.

The Analysis

Figs 3 to 6 contain the stopwatch time subtractions, and fig 3 shows the difference between the real time (E) and the Strike time (S). It can immediately be seen that there are two broad groups of data – at about 9.7 and about 8.4 seconds, with probabilities of 1/3 and 2/3 respectively, and with a total spread of about 2.1 seconds

Fig 4 shows how long the hourly Westminster Chimes take – a fairly consistent 17 seconds, but admittedly with two rogue times, at 18.7 and 15.9 seconds. Two readings in twenty is a probability of 10% that the Chimes will not be consistently 17 seconds!

Fig 5 shows that there exists two broad groups of data at about 26.8 and 25.4 seconds, with one rogue of 25.9 seconds. This time is that which separates the real time (E) from the Quarter Cam lever release time (Q), indicating that there is a degree of slack in the movement here.

Conclusion

Fig 6 shows that slack is confirmed between the Hour and Quarter cams, which are both secured on the one arbor, and which show two groups of data at 29.9 and 28.4 seconds (the time difference of 1.5 seconds is due to the pendulum drive of course). It must therefore be that this arbor flip-flops from the one leading gear tooth to the trailing one – a very small gap between the gear teeth which allows either one contact or the other, but nothing in between. This then is the cause of the Strike Bell time scatter.

The going train is driven by a weight – this is not applied to the "H and Q cams and hands" arbor, but upstream near the escapement arbor. Thus the weather is an important contributor to the Strike Bell time accuracy because wind drag on the three clock face hands can cause the H and Q cams arbor to flip-flop from one side of a gear tooth to the other.

The clock is now rewound electrically, and drives closer to the escapement than when hand wound, but the (H and Q cams and hands) arbor has always been exposed to the forces of the wind.

Stroud Parish Clock of St Laurence

Stroud Parish Clock Strike and Quarters cams

Stroud Parish Clock Quarter Chimes Drum

Restoration

Over the years I have restored several clocks: The Stroud Parish Clock c.1878 at St Lawrence as above, and the Uplands Church Clock, Rodborough Church Clock c.1887, Bownham Park Clock c.1908, Nailsworth Central Pillar Clock c.1851, Woodchester Church Clock c.1876, and Coaley Church Clock c.1720 which I helped Michael Maltin DFC restore in his garage.

In August 2006, I was cleaning the clock from Coaley Church, with Michael Maltin DFC, which we think was made in 1720 – Sir Isaac Newton died in 1727 and Harrison's H4 came along thirty years later - see ISBN 0-00-771883-7 "Longitude". The 1720 date for the Coaley clock comes from a meeting of the Antiquarian Horological Society to see the local turret clocks, of which this was one. The general consensus was circa 1720. The clock is of particular interest to me, because it shows the engineering design and make skills they had in those days. I find myself wanting to know much more about their skills – school and TV teach about the development of science and mathematics, but not so much about engineering design and make. I did not realise, for example, that steel was made in Roman times by blacksmiths forging and folding the iron they worked with in small quantities each time – this reduced the carbon content of cast iron (by emitting sparks) to make steel, but they had to judge through experience how many times to fold and reheat it to get the best steel. The Henry Bessemer furnace in 1858 was the first to make steel in large quantity, but only after the process was changed by metallurgist Robert Mushet by adding 1% carbon back to a well oxidised molten carbon free iron.

Those were the days when every nut and bolt was made individually and without particular standardisation – Joseph Whitworth was the first to standardise threads throughout the UK in 1841, over 100 years later. So, when removing bolts, I had been carefully noting in which position they were fitted, so they could go back where they came from!

It was interesting to see the rounded crests of the screw thread itself, and I wonder if that was due to their manufacture – rolled bolt threads perhaps. Using a strong magnifying glass seemed to indicate rolling, but there were very small cavities where the top of the steel thread form is missing, and at the moment I do not know why. I found that the nuts would fit onto several of the bolts, so local standardisation of thread form was evident in that particular foundry.

The 1720 clock had been recently found in a wooden shed in Coaley Church grounds, overgrown with ivy – the shed had stored several lawn mowers over the years and the donation of yet another mower to the church may have triggered a clear out! The local Museum was contacted by the church and thus they phoned Michael Maltin, who, with his long background in restoration, was asked about it – he said it was valuable for its time, and we are now preparing it for the Stroud Museum in Stratford Park, since they have no ability to do so.

It may have been on its side for many years, since this 1720 clock was replaced by another and that in turn replaced by the current clock in 1939 I think. One of the arbors which protruded beyond the 2.5 ft x 2 ft x 1.5 ft frame was bent and that can only have been due to a heavy weight landing on it – the clock in operating condition would not be able to do that, except if it had been dropped on its side. The local forge have pretty well straightened it for us.

The wear on the clock is wonderfully descriptive of the clock mechanism – worn dimples on the bell strike wheel and the lever which produced those dimples very much worn itself with a deep Vee groove. The bell was struck the same number of times as the hour itself – the clock may well have driven dial hands, but some of the lead-off parts are missing.

During the strip, I have also found an old hand-manufactured split pin – 1720 or a later fitment I do not know, but the skills involved were clearly good. The gear teeth are excellent – all well spaced and formed, and a good late repair on the escapement wheel also evident.

The escapement is of the old anchor recoil type – the second hand, if there was one, would have moved forward then slightly back (recoil) before going onto the next.

Calculations of gear ratio suggest that the pendulum had a 1.5 second beat, requiring a pendulum length of 7 feet 4 inches (2.24 m), though that is now missing.

Having wire brushed the clock parts, I have rust proofed it with a modern rust treatment product (Trustan), which gives a satisfying black finish and which does not obscure any manufacturing marks – painting it would not be good because it tends to obliterate everything underneath it.

Note: The c.1600 fieldgate clock in Leonard Stanley church which Michael restored many years ago, did not have a dial (merely striking the hours on a bell – note: the Medieval Latin for bell is clogga, and thus the origin of the word "clock") and predated the pendulum which was first used in a clock in c.1657, although "Old Solomon" was modified to have a pendulum, many years before the restoration.

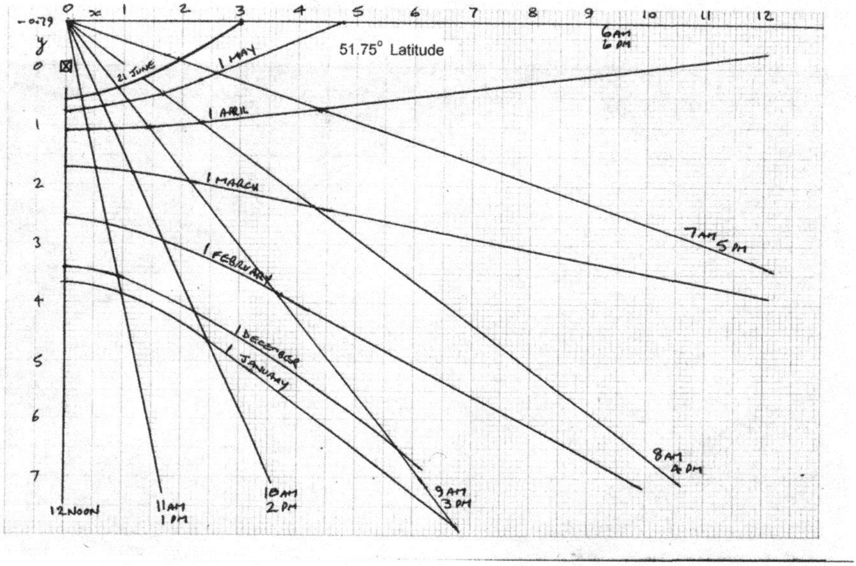

SUNDIALS

I joined the BSS (British Sundial Society) to enjoy the September meetings in Newbury with Michael Maltin and Vera Naumanns, and saw many different types of sundial there, but none were of a simple post, so I spent time researching how to design a sundial and I used my computer to design the placard on which the pole stood using the equations in ISBN 085950-062-4 "Positional Astronomy and Astro-Navigation made easy", and produced the above chart, that I presented to the BSS.

Note that I first plotted the positions on the chosen date, and then discovered that the times were straight lines when plotted through those date lines. Note also that the focus of the straight lines is North of the post by 0.79 times the height of the post. So, if making this sundial for real, you might make a post 2 metres tall, and you would need your lines focussed at 1.58 metres to the North of the Post. Here is my small Version, showing the time is 4.35:

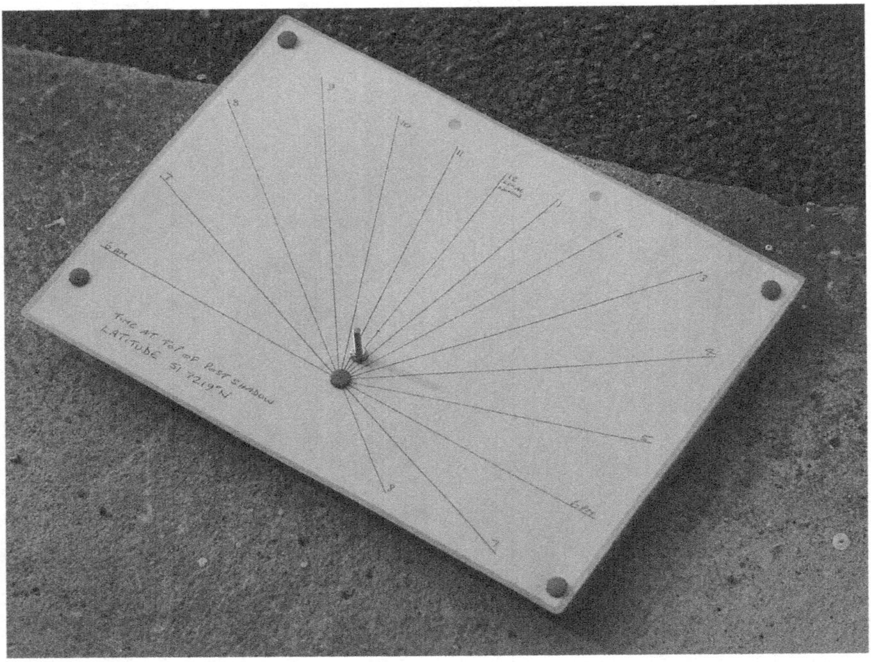

CHAPTER 16

OLYMPICS ORGANISATION

Considerations I proposed in 2007 to LOCOG for the 2012 Olympic Games Organisation because I wanted a job with them were:-

a) The IOC Website

- Does not define Sports Rules
- Rules lists not even available as a booklet
  FIA Yellowbook equivalent not available
  How many athletes allowed – how many heats?
- So, Beijing Games are to Chinese Rules?
- How do Britain's Rules differ?
- When an athlete appeals, whose rules?
- IOC have appeals procedure, but no rules!
- Unfortunately, cannot find Rules on BOA, NOC or FIG websites either!
  Eg What dimensions and how high should Gymnasts rings be?

b) Rules Example

- 100m track, 10 seconds, 0.001s timing accuracy = 10mm distance
- Measurement accuracy: 100m steel tape expands 6mm through 5 degC change, so how?

- Parallelism of start and finish – accuracy not defined. Swimming pool too.
- Start and finish lines are very wide! Compensate: fix timing lights (height) accurately.

IOC are going to need calibration dimensions for all events. These checks need to be in timescales programme.

c) Typical Routing and Computer Displays based on my car Rally Stage procedure:

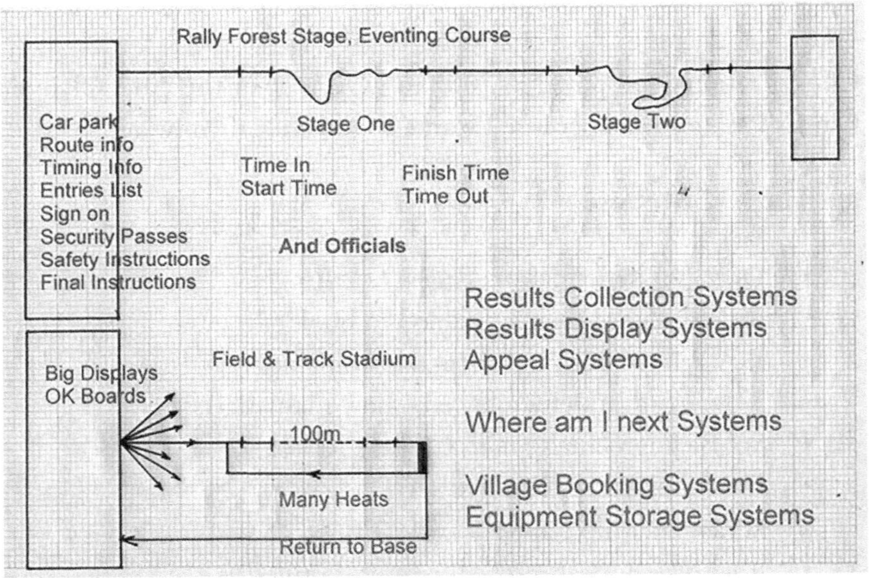

d) Venue Planning

- Need Computer Displays Outside Venues for crowd control if evacuation needed
- Need redundancy built into Event Timekeeping Planning to allow for bomb scares holding up event
- Need Car Ferry Route Planning + RAC Route Planners around venues – road closures, speed limits, one way streets – RAC/

AA road signs – are car parks big enough (eg media people regularly moving around Park).

- Need bus and rail Regularity and Timekeeping agreed, including Eurostar
- Need whole event planning to recognise London rush hour – or change rush hour! One way streets to change direction morning and evening. Speed limits to be painted on roads, including 30mph, for foreign visitors especially.
- Need Competitor Liaison Officers to organise competitors, answer their questions etc
- Need fire trucks and other rescue equipment, properly manned and planned
- Need doctors, nurses and ambulances for competitors – and spectators too
- Need to plan Olympic Torch and Marathon Routes.
- Need clean food and hygienic toilet regularly cleaned, male and female separate.
- Need Venue Safety Planners eg keep people away from archery targets
- Need bright arrows for competitors
- Different colour Signposts for spectators
- Need maintenance crews – electrical, mechanical, software etc
- Need turnstile money collectors – and collectors from them. "Tickets-before" collection too.

e) Timescales for 2012 Olympics

- Tests must include all facilities in Summer 2011
- Olympic Park Delivery Programme shows March 2011 for Tests Start, but Village too late? (December 2011).
- Roads, trains, buses timetables planning needs at least a year, and these must be in place at start of tests too – so 2008 to launch final planning – Displays for travelling public? Strategy planning – 2007.

- Computer Systems need recruitment to start in 2008, but Requirements Strategy needs Planning 2007.
- Consult RAF, RN, Army, Police, Fire, Ambulance – 2007.
- Publish Olympic Competitors Regulations 2011 – need an IOC example set.
- IOC are going to need calibration dimensions for all events (2012)

f) Beijing Olympics 2008

Who should go?

- Chairman of London Olympics Board
- Athletes!! When they return, what did they like, not like? Handover all their paperwork to us and a few take us through it, from initial application to goodbye, for us to think about London. They need these thoughts too, so that they know what to expect before they go.
- Who else NEEDS to go?

Manchester history should provide good launch info! And India's Commonwealth Games in 2010 also of interest.

CHAPTER 17

THE EUROPEAN UNION

The UK's net contribution of £6 billion per year into the EU in 2014, helps (together with Deutschland, France, Nederland and Italia particularly) to provide countries like Poland and Hungary with money to recover from their desperate years under the Soviet Union – I support this provision, which is a net contribution to the EU of £115 million a week. I think we should name countries and cities by the name they call themselves! I support the EU because of the hiatus that occurred after the First World War when so many European Nations were territorially carved up, and yet the Second World War occurred as a result of so many Empires being destroyed militarily after the First World War, and which gave rise to so much nationalistic anger. The European Union was thus formed to trade between the Nations and to avoid war, and for this reason I support it, albeit it has a few problems of organisation to sort out.

The contributions to former Soviet Nations will reduce their people's need to migrate towards those rich Nations like the UK.

One of the reasons that Brexit happened because of the Leave campaigns strength, was that the EU Council and the EU Parliament are not sufficiently democratic, but I worked well with MTU Munchen and Fiat Torino Engineers before the UK joined the EU, so I strongly support the idea of working with Europeans. I think that leaving the EU will actually diminish the UK's standing on the world stage, because the big nations of the world will listen to Europe in preference to the UK, and the UK could have influenced the EU far more if it was properly committed inside the EU.

We had a chance in the EU of keeping English as the main alternative language, but they may now promote the French language instead – holiday makers today may get away with English while in France, but that is because the French learn English at School. This will possibly change to Deutsch after Brexit? I have noticed that non-UK MEP in the European Parliament often make speeches in English, presumably because they communicate better to a wider audience on their subject, without needing translators and bulky earpieces. I have also heard that it is often quicker to communicate an argument in English than it does in French. I wonder how long it takes in Chinese with their very complicated looking symbols for words? I have bought ISBN 978-1979789080 "50 Shapes of Chinese". I have also watched Chinese people talking on TV with subtitles – they speak very quickly compared with the English subtitles – so reading or writing a newspaper might be slow, but talking to each other about the news would be as fast as we are probably.

I voted REMAIN (to stay in the EU) much because I wanted to preserve the peace that has existed in Northern Ireland following the Good Friday Agreement in 1998, and which the EU system naturally supports. I also think they are good trading partners and have good food, cosmetic and pharmaceutical standards.

I observe that the EU Council gives each country the same size, or weight, of vote, whether big or small, but that Qualified (being one of the big 6 countries and in which Malta for example does not have a vote therefore) Majority voting mostly occurs by the big six countries. No wonder Theresa May had difficulty making sure her voice was heard, even before she was Prime Minister – the UK has been demoted by the EU politically into a small country for a non democratic vote. Surely each Minister representing their own country, should have a vote proportional to their countries population. Thus the UK should have 65,140,000 votes and Malta only 429,000. However, I note that the European Council meets quarterly and 80% of votes are by Qualified Majority, which means that 65% of the population of the EU have to be represented by appropriate Member State Ministers – this makes the bigger States vote count for more. I have written to the President of the EU Council, about this lack of a balanced democracy both in the EU Council and the EU Parliament, and I have received an acknowledgement of my letter which says that the British Government agreed the current position! So, it must be that the UK wants the smaller countries to have a bigger shout when it comes to voting.

The EU Parliament also has big differences between the different countries – the UK has one MEP for every 833,000 people, Malta one MEP for every 84,000 people, and Luxembourg one MEP for every 97,350 people.

The EU reports the facts at
https://ec.europa.eu/trade/policy/eu-position-in-world-trade/index_en.htm#

I see that Trade in Goods and Commercial Services in 2013, in euro billions is:-

Country	Imports	Exports
EU	2188	2415
USA	2079	1688
China	1716	1817
Japan	750	648
South Korea	468	506

EU trade is 16% of the world market, leaving 84% in the rest of the world, but the EU has retained its capacity to conclude and implement even more trade agreements. The UK has to recruit staff and start trade agreements all over again, from now. Surely we will have to pull out of all of those EU Trade Agreements so far made, and initiate our own agreements with those same nations. It should be noted that UK-EU trade relative to the rest of the world will reduce because the rest of the world is growing so much, but our trade with the EU will actually grow in £pound terms – the "leavers" scaremonger by saying that EU trade will reduce, but they are talking of %ages which is irrelevant here, for they clearly have not understood what trade is all about - £pounds not %ages!

The European Union has been good to us for over 40 years. We have good Laws over that period and we have good workers rights. The EU has brought many benefits to airline passengers in Europe by liberalising markets, improving safety and security, and protecting consumers and promoting International business. The Single European Sky might not be a UK idea, except for the fact that we are in the EU today, but it is strategically necessary nevertheless. We are going to have to be very actively engaged with all of the aviation committees after Brexit, to make sure our views are clearly put.

We have a human rights act that has stood the test of time – any arguments should be placed with the EU, not create our own version – we must involve and commit ourselves more in the EU, and not give up if we have a good argument. The UK has half of its immigrants coming from inside the EU, and the other half from outside the EU, but non-EU migrants to the UK are not bound by EU Rules as Mrs Theresa May and the Home Office appear to have said. Therefore the UK could have severely reduced its non-EU migrant input.

The claim that we lose £350 million a week to the EU is just plain wrong, for the actual net contribution figure is less than half that, because of the support that the EU gives back to us, recognising our achievements and ability. After Brexit, the UK Government should support all those who would otherwise have received funds from the EU, which means the NHS will not get anywhere near half as much as the "leave" campaigners said.

China is a big nation that is now developing particularly good aircraft (the J10, J15 and J20) and other weapons systems. We cannot compete with them on a war footing at all for their numbers are so huge. Our new aircraft carriers will not be anything other than a pin prick in their side. The same goes with the USA, and Russia. We will be a world power aided by staying in Europe and welcoming their solidarity and assistance. We have to get over this arrogance that we are better than Europe can be. We did not win the last war, the USA and Commonwealth countries citizens provided far more manpower than we did. That has to be recognised and the UK stop thinking it did it all itself. I think also the UK does not give enough recognition and regard for the incredible bravery and support to winning that the French Resistance showed during the second world war, and I find that very disappointing. However, I am pleased the French Army were allowed to lead the liberation march into Paris in August 1944, for it recognised the incredible bravery and support of the French Resistance too, and even though obviously all of the Allies were necessary to oust the Deutsch in the first place.

The euro has suffered some troubles over the years, because the individual nations were not being helped to see how to control their own finances within it – this was the EU Commissions fault for not taking a helpful attitude and painting a picture of what it was all about. Individual nations spent too much, thinking they had the euro to protect them. It was not the euros fault, but the way it was handled. They are learning to come to terms with it. We surely stay out of it.

Mean time our foreign exchange rate dived and many UK people living in the EU suffered badly from the drop in the value of the £ pound back in 2007 or so, and it is now 35% less than it was in 2004. This was rarely mentioned by the media, but it shows the UK has suffered financial problems too. The £pound has suffered as a result of the EU Referendum vote, and we have survived it so far, but nobody has mentioned the big financial effect it has had on our people living in the EU. These people are often pensioners who only receive money from the Government in £pound terms not euro so they lose out when the £pound drops.

The EU have many negotiators for they are used to trade barrier talks, while the UK has lost most of its expertise here. We are going to have to start trade talks all over again with a new and inexperienced team of civil servants, due to that EU Referendum result. The UK lost its trade team because it did not think well enough at Government level. The EU have been negotiating trade talks with other nations outside the EU, and we should have been far more strongly represented there if we want better trade with other nations.

TRADE AGREEMENTS

I decided to investigate a few EU Agreements, to help make up my mind about the UK Governments Brexit Policy on Trade Agreements:

1) KOREA

I have briefly looked at the 1432 page EU document that is the EU-Republic of Korea Free Trade Agreement - it is contained in the Official Journal of the European Union Volume 54 dated 14May2011. On page 81 is shown the signature of The United Kingdom of Great Britain and Northern Ireland. In its preamble it restates the good principles that I like, which I select as follows:

REAFFIRMING their commitment to the Charter of the United Nations signed in San Francisco on 26 June 1945 and the Universal Declaration of Human Rights adopted by the General Assembly of the United Nations on 10 December 1948;

RESOLVED to promote transparency as regards all relevant interested parties, including the private sector and civil society organisations;

DESIRING to raise living standards, promote economic growth and stability, create new employment opportunities and improve the general welfare by liberalising and expanding mutual trade and investment;

DESIRING to strengthen the development and enforcement of labour and environmental laws and policies, promote basic workers' rights and sustainable development and implement this Agreement in a manner consistent with these objectives;

BUILDING on their respective rights and obligations under the Marrakesh Agreement Establishing the World Trade Organisation, done on 15 April 1994 (hereinafter referred to as the 'WTO Agreement') and other multilateral, regional and bilateral agreements and arrangements to which they are party;

and in Article 1.1:

(c) to promote competition in their economies, particularly as it relates to economic relations between the Parties;

(d) to further liberalise, on a mutual basis, the government procurement markets of the Parties;

(e) to adequately and effectively protect intellectual property rights;

(h) to promote foreign direct investment without lowering or reducing environmental, labour or occupational health and safety standards in the application and enforcement of environmental and labour laws of the Parties.

"South Korea offers many opportunities for British businesses in sectors including retail, clean energy, fashion, industrial design, automotive, and ICT."

2) THE COTONOU AGREEMENT

The Cotonou Agreement amended in 2010 and signed by all 27 EU Member States (including "For her Majesty The Queen of the United Kingdom and Northern Ireland") is a Partnership Agreement through to the year 2020, with 75 countries which were party to the Georgetown Agreement of African, Caribbean and Pacific States (ACP). The 224 page "The Cotonou Agreement" (in part "AFFIRMING their commitment to work together towards the achievement of the objectives of poverty eradication, sustainable development and the gradual integration of the ACP countries into the world economy") is an agreement by which the EU Member States and the ACP States collaborate on Institutions, Cooperation (Development, Economic and Trade), and Development Finance. It recognises that the ACP States need time for their markets to be liberalised. It also says in Article 36: "1. In view of the objectives and principles set out above, the Parties agree to take all the necessary measures to ensure the conclusion of new WTO-compatible Economic Partnership Agreements, removing progressively barriers to trade between them and enhancing cooperation in all areas relevant to trade."

With The Cotonou Agreement's level of careful thought, I am very concerned by what the UK Government thinks it is going to do with the ACP States either individually or collectively after Brexit. It is not enough to say that the UK is going to create Free Trade Agreements with any of these ACP States, for that could make these ACP States accept great risk to themselves – the UK must be cautious here, for I think the EU has already set the Good Standard, and which will no doubt be carefully amended again for 2020.

3) CHINA

There is no Agreement with China, but I have read and fully support the 19 page "Elements for a New Strategy on China" by the EU Commission dated 22nd June 2016, which provides a Policy Transparency that is missing from the UK Government. It shows how much the EU has been thinking about China's role in the world, including their Human Rights record (poor IPR Intellectual Property Rights are a further example, but steel overcapacity has been somewhat reduced by China, and the EU have used an "analogue country" model to predict China's costs from which State subsidy might be determined and EU tariffs might be proposed – anti-dumping tariffs by the EU on China exist). Clearly the UK will have little strength here if it leaves the EU, and trade would be on China's terms, not the UK's nor the EU's. If the UK negotiates free trade with China, it will be because we ignore everything unacceptable that exists there – this would also arrogantly presume our strength relative to the EU who have been pursuing a principled trade agreement with China for many years, undoubtedly with UK support.

The UK's best prospects for a favourable trade agreement with China, a strong currency and rules-based trade, are to be found by remaining within the EU. The EU has long been reluctant to grant China market economy status until it can demonstrate that Chinese product prices reflect their market value and without subsidy therefore. The EU has sought to develop a more cohesive approach to Chinese investment in EU countries. In holding China to rules-based trade, the EU is following the same approach that Brexiteers claim to. So, I do not want the UK Government Brexiteers to pursue a FTA without using EU Principles.

It should be remembered that Deutschland exports far more to China than the UK does, so one has to conclude that the UK lacks the incentive to trade and does not have the right goods to sell, and separating from the EU will not help that. The February 2018 trip to Beijing during which Theresa May announced that the UK Business Delegation had made £9billion of deals was small by comparison with that that Xi Jinping made when he came to the UK in 2015. These were not UK Government but UK Business deals in spite of no trade agreements, and I am concerned that the UK will adopt poorer principles than the EU have in their "Elements for a New Strategy on China" on trade agreements. China is far more interested in the EU than in the Brexit UK in any case.

4) USA

I am even more worried by any agreement that the UK might make with the USA. TTIP (Transatlantic Trade and Investment Partnership) was at one time being secretly negotiated between the USA and the EU – perhaps it is on pause at the moment thank goodness, but I could find no EU Document about it, merely media reports. I hope the EU stays principled in its discussions with the USA. It opens up a quagmire of aggressive USA Businesses wanting to force their way into the EU with poorer workers rights and the opposite of the EU's substance Law – for example the EU currently bans 1200 substances from use in cosmetics because they are proven to be a health risk, as against just 12 in the USA where they think that any substance is safe until proven otherwise through population exposure. Even Britain's NHS is at risk here. Clearly however, EU goods are qualified to a better level than USA goods, so we can export to the USA and meet their Regulation Requirements, but the EU cannot accept USA goods because they do not meet our EU Requirements. The media have also warned us about chlorinated chickens, GM foods, hormone beef and low quality milk coming from the USA, and yet UK Government Brexiteer Ministers are prepared to accept this awful state of affairs. NO to TTIP with these repugnant USA rules – NO to a UK-USA trade agreement too, dominated as it would appear to be by the USA.

5) INDIA

I have looked at the chance of negotiating an EU-India Free Trade Agreement – ref the EU Parliament's report dated October 2013 "Workshop The Long Road towards an EU – India Free Trade Agreement". I decided to look at the EU-India potential trade agreement instead of EU-South America (the incomplete EU-Mercosur Association Agreement). The EU-India agreement has been stalled since 2013 because the Indian Parliament cannot agree FTA on all goods and services, and in general it's tariffs are higher than those of the EU – in particular India wants tariffs to protect domestic manufacturing industries, and it's Ministry of Finance does benefit from a big source of revenue. If one looks at the chemicals or car parts sectors, the Indian tariffs were 7.5-10 percent in most areas. The wine and spirits industry is a highly protected sector in India, with 150 percent tariffs on European wine and spirits products. Tariffs also exist on other key sectors such as cars, where tariffs were at 60-100 percent in 2013. Also, without key outcomes on sectors such as banking, and insurance, it will not be possible to conclude the negotiations.

IPR is not a core issue in the negotiations and the EU does not require India commitments to modify its domestic legislation in this regard. On the 2013 environment of forced child labour, India has legislation in the Indian Parliament, which if adopted and enforced, would allow India to ratify the ILO Conventions on Child Labour, thus possibly satisfying the EU, but this seems even further off than it was five years ago, and no agreement with the EU will result until child labour in India is totally abandoned. In addition to a fear of competition, there is also a fear of getting too many manufacturing goods from the EU so as to cause a large current account deficit. Further, there is a major shift in India's negotiation strategy and pattern to the East (that is, towards the ten Member States in ASEAN – Association of South East Asia Nations). The Indians want better Visa arrangements for their citizens going to the EU.

With respect to investment, India has signed bilateral agreements with different EU member-states, and the problem here is that no one yet knows which one should be the model incorporated into the EU-India FTA agreement. It will not be possible to have a Trade Agreement similar to the EU-South Korea one. In 2007, the EU and India began to negotiate a Broad-based Trade and Investment Agreement (BTIA), and maybe this is still the best way forward. This will be India's first trade agreement with a developed country and its most comprehensive trade agreement. It is also the EU's first trade agreement with a large emerging market. It seems to me that the UK Brexiteers are not going to negotiate any more of a principled Trade Agreement with India, than the EU is still attempting.

6) WTO and WCO

The World Trade Organization (WTO) deals with the global rules of trade between nations. Its main function is to ensure that trade flows as smoothly, predictably and freely as possible. The WTO World trade Organisation and WCO World Customs Organisation are intergovernmental bodies that regulate how Trade and Customs are to be organised with an emphasis on reducing or eliminating trade barriers between States, but they do not at all prescribe the tariffs that any one State applies to another's goods and services – they merely record them. The WTO represented 164 Member States in July 2016. The WCO in December 2017 represented 182 Customs administrations that collectively process approximately 98% of world trade. Thus the UK Brexiteers who talk about reverting to WTO rules do not know what they are talking about, for all UK trade is already in compliance with WTO and WCO rules. See for example WT/L/940 dated 28th November 2014 for 32 pages of "Protocol Amending the Marrakesh Agreement Establishing the WTO" which includes guidance for developing country and least-developed country Members. The following detailed webpage lists those tariffs imposed by States, which have been notified to the WTO and thus collated, but they are not imposed by the WTO at all. Some bigger EU and UK Tariff Examples: 20.8% on frozen strawberries, 25% on prepared pineapples, 33.6% on unfermented fruit and vegetable juice, 32% on wine of fresh grapes, 74.9% on tobacco.

The WTO Agriculture Agreement has determined that the maximum level of subsidy that developed countries like the EU can apply has to be reduced by 20 percent over 6 years.

So, it all depends on what tariffs and subsidies the UK Government finally sets as a result of Brexit, and what tariffs it has to put up with from other States (including the EU) - it cannot expect free trade with anyone just because it asks for it.

Some Brexiteer MP say that if we do not have a FTA with EU we will have to resort to WTO Rules, implying that *de minimis* rules are actually legally binding on States – they are not, so Brexiteer MP are perhaps ignorant of the WTO.

de minimis: The WTO Glossary says: Minimal amounts of domestic support that are allowed even though they distort trade — up to 5% of the value of production for developed countries, 10% for developing. From Latin: de minimis non curat lex 'the law is not concerned with trivial matters'.

It is noticeable that the EU and the UK have very many tariffs that are smaller than 5% - these could be called *de minimis*. The EU and the UK have many tariffs that are above the *de minimis* threshold too. *De minimis* means anything under a small value. The WTO rules mean that it will not persuade a State (through economic argument) to reduce any of its tariffs that are *de minimis*. It does not impose a *de minimis* tariff on any State, for States are free to set their own tariffs. A Free Trade Agreement can have *de minimis* tariffs, but tariffs of any sort whether *de minimis* or not, are not "WTO Rules".

7) WTO ANTI-DUMPING AGREEMENT

If a company exports a product at a price lower than the price it normally charges on its own home market, it is said to be "dumping" the product. The WTO Agreement does not regulate the actions of companies engaged in "dumping". Its focus is on how governments can or cannot react to dumping — it disciplines anti-dumping actions, and it is often called the "Anti-dumping Agreement". The WTO Anti-Dumping Agreement allows governments to act against dumping where there is genuine ("material") injury to the competing domestic industry. In order to do that the government has to be able to show that dumping is taking place, calculate the extent of dumping (how much lower the export price is compared to the exporter's home market price – the EU uses their "analogue country" method – for example, Brazil was chosen for footwear to compare with Chinese and Vietnamese prices), and show that the dumping is causing injury or threatening to do so. Calculating the extent of dumping on a product is not enough. Anti-dumping measures can only be applied if the dumping is hurting the industry in the importing country. Anti-dumping measures must expire five years after the date of imposition, unless an investigation shows that ending the measure would lead to injury.

Anti-dumping investigations are to end immediately in cases where the authorities determine that the margin of dumping is insignificantly small (defined as less than 2% of the export price of the product).

Typically anti-dumping action means charging extra import duty on the particular product from the particular exporting country in order to bring its price closer to the "normal value" or to remove the injury to domestic industry in the importing country.

Anti-Dumping measures taken by the EU (23october2017): against Belarus, China, India, Republic of Korea, Iran, Russia, Serbia, Chinese Taipei, Turkey, Ukraine, Malaysia, Thailand, Tunisia, USA which are all under review or tariffs are now set.

It should be noted that USA Senators, for example, can be Senators for all their lives (minus 30 years for they have to be that old to be a Senator), so long term experience might make "sensible voices" in USA tariffs!

8) CONCLUSION

It is quite clear that Trade with the Developing World is recognisable on the EU agenda. World Trade outside the EU is set to grow considerably, but the UK can Trade with whoever outside the EU it wishes to Trade with while being an EU Member State, and to say otherwise is yet another Brexit lie. The Brexiteers attempts to say that to take advantage of the expected growth of World Trade means the UK has to be outside the EU is a nonsense. We export Military Equipment to Saudi Arabia big time, and we do not have a FTA with them, so being inside the EU was obviously not the hindrance the Brexiteers have said.

I have decided not to dig any further into trade agreements. I am happy that the EU is maintaining its principles in all of the above trade agreement negotiations, but it leaves me concerned about the UK Brexiteers arguments - what principles will the UK Government have after Brexit in any of its new trade agreements? It is obvious that Brexit causes a need to replace all of the EU Trade Agreements as far as possible, but surely we have to be principled in all our agreements, not just claim how much better off the UK is outside the EU.

I think that the racist remark made about US President Obama's family origins and heritage during the EU referendum campaign was "out of order", and I think The President's opinion at that time that the UK should remain in the EU was rational and "in order", but that rejection of his opinion was wrong I think. It is this Brexiteer attitude that make me wonder about the UK Governments Principles in New Trade Agreements.

CHAPTER 18

GOVERNMENT

1) I proposed a Safety Strategy to the Deputy Prime Minister for Britain's Railways following the 5th October 1999 Ladbroke Grove crash – fundamentally, I thought we needed coloured railway traffic light signals to be repeated on a miniature display in the Drivers Cab. This was missed in the Inquiry by Lord Cullen. Signalmen were reported by Lord Cullen to work over 12 hours a day – that seems long to me for a responsible safety job, but he made no such comment. His comment was about the weekly 72 hours limit and that men should not have been authorised to go beyond it, but 72 hours is surely too long too. He was a lawyer used to Inquiry, but was not an engineer.

2) I have supported Baroness Hamwee (she was kind enough to reply to me several times) in the Guardianship (Missing Persons) Bill, now given the Royal Assent and is now therefore an Act, but she tells me it might be a year or so before the Act actually becomes law. This is because there are necessary regulations and orders, she says, which will take that long to prepare. The Secretary of State would have to raise an SI to be agreed by both Houses of Parliament to instruct those regulations and orders. I found a shortcoming in the Bill, by examining it logically, like any interested software writer could. She did not accept my suggestion because she wanted the Bill passed before the 8th June 2017 General Election, and I supported that too. Bills apparently are dropped when a new Government comes in, and she did not want that, and nor did I, for there have been reports in the media about a missing persons estate having been decimated by inconsiderate bankers who would not allow that persons family to control the bank account. I did not see the Bill early enough to put my amendment in to her in sufficient time, which I did on 22nd April 2017. The third reading in the House of Lords was on 27th April 2017 and was passed, so it was given the Royal Assent that day by the Lords. The shortcoming I found was real in that my sister and her husband live in Canada, but her son lives in the UK. If my sister went missing the Act would not allow her husband nor her son to control her estate in the UK, and I think the Act should be amended to allow that – she has not gone missing fortunately!

3) I have challenged the Governments Spaceport Bill in a letter to the Prime Minister, for it seems to have ignored the fact that there will have to be several square miles of protected land (where would you find that in the UK?) where people will not be allowed during a rocket launch, because its white noise can so badly damage hearing – for evidence look at the miles distance from take-off that Florida people have to be, on TV pictures. They also appear to have forgotten that the downstream distance to possible failure could lead to casualties over Norway and Denmark which are only 6 degrees East (250 miles distant) and therefore not far away from the launch point on the East coast of Scotland, which is the only possible site that I can envisage. The NASA and ESA launch sites have the whole Atlantic to contain their possible rocket failures, the UK does not. Therefore if the UK were to launch spacecraft it will have to launch into Polar Orbits, at least until sufficient reliability is demonstrated to allow Equatorial Orbits over Denmark and Norway, if ever that was possible.

4) AVIATION

ICAO (International Civil Aviation Organisation) identifies 52 member States in the European Region, while EUROCONTROL identifies 41, the EU is 28, and NATO 26, all including the UK.

The EU has brought many benefits to airline passengers in Europe by liberalising markets, improving safety and security, and protecting consumers. Brexit is therefore disappointing here. Air traffic is expected to double in year 2030 from what it was in 2009.

Air travel is international in nature and thus has to work across National Boundaries, hence the UK was one of the founding six States in EUROCONTROL in 1961, during the Cold War. Unfortunately EUROCONTROL found it was not so well able to Regulate as it wanted to, so delegated that activity to the EU Commission. EUROCONTROL is the pan-European technical support and management arm (eg development of implementing rules, of specifications, performance of PRB tasks - Performance Review Body).

The SES (Single European Sky) project for ATM (Air Traffic Management – ATC: Air Traffic Control) has thus evolved since 2004 into an EU Regulated activity, which may no longer be supported by the UK after Brexit, but in which the UK played a leading role. I think that the UK should continue to be a Member State of EUROCONTROL, to introduce the technology needed across Europe and for our own Industry to create and support. SESAR is a part of that technology introduction plan, which I think is similar to the FAA's NextGen system, and which is being implemented by the UK NATS (National Air Traffic Services) since 2015 and on into 2019. It is estimated that European ATC's will require perhaps 15% more staff over the SES introduction phase. NextGen and SESAR are going to make more use of satellites and digital systems replacing the old VHF systems – SESAR 2020 was launched by EUROCONTROL in 2016.

Note that NATS Holdings Ltd is a privatised Company whose accounts can be seen on the Companies House website, while the CAA (Civil Aviation Authority) is a government subsidiary not a private Company. After Brexit, NATS will require the CAA to provide the Economic Regulation that the EU currently does. More than that, I think the UK should continue to be a Member State of ICAO's European Region, and look to involving Russia and the former Soviet Union States into SESAR deployment as well. The SES initiative should be more embracing of those additional Nation States too, and thus why I think EUROCONTROL should develop better relations with Russia, embracing their needs too, but maybe ICAO Europe Office in Paris does that already. After Brexit, the RAF will need access to the Single Sky Committee to ensure Military flying rights. RPA (Remotely Piloted Aircraft) will also need integration into the SES. EUROCONTROL used UK Met Office charts to coordinate responses to the Icelandic volcanic ash crises of 2010 – ICAO setup a committee to resolve how to better improve response to such ash clouds.

The provision of ANSP (Air Navigation Service Provider) will always remain a public task – ref ISBN 978-90-411-3730-2 "Achieving the Single European Sky". The average size of an ANSP in Europe is well below the average size in the USA – 50 to 16 respectively which leads to more cost and more delays and emissions – aircraft also zigzag to avoid military areas. We need to ensure the experts get together and agree ways forward under the general directive to reduce the number of ANSP. This goes some way towards explaining that the cost in the EU is some 60% higher than in the USA. The UK NATS is an ANSP, and the CAA is the National Supervisory Authority, and both should remain so.

A problem that needs to be resolved since the EU Referendum in the UK in 2016, is the prior introduction of the UK-Ireland FAB (Functional Airspace Block) ref https://www.ukirelandfab.eu/about/, which is the oldest FAB in Europe, but I hear that UK Air Traffic

Controller Staff think the EU Regulations are too burdensome here. Nevertheless, I support the logic of the FAB system for the improvement of ATM across Europe. The very purpose of FAB's is to reorganise the fragmented airspace into large airspace blocks where sectors are delineated according to operational and technical needs, instead of national boundaries.

The CAA CEO has published a Paper in December 2016 in which he outlines four important Principles for UK Government Policy Enactment after Brexit.

I support all of these, but I am very concerned by the CAA's document on their website dated 2014 "State Safety Programme for the United Kingdom" wherein it is clear that the EU Regulations provide the overarching framework for the UK – we are totally integrated in aviation with the EU and EASA here (European Aviation Safety Agency). I do not know how the EU Regulations come about – to what extent do the UK Department for Transport and the CAA provide intimate guidance to the EU? Nevertheless, it is clear that we will have to negotiate with the EU over the Regulations, or we will not necessarily comply. So, we do need a strong voice with the EU, not at all to ignore them. We told ICAO in 2003 that EASA was the UK Government's authorised agent for the fulfilment of specified State of Design and State of Manufacture obligations under ICAO Annex 8, and EASA has evolved since. We will have to decide whether we are going to continue delegating to EASA, but I think we have to.

The CAA's "Future Airspace Strategy for the UK from 2011 to 2030" is dated 30June 2011. The FAS (Future Airspace Strategy) Vision is to evolve and identify what needs to be done between 2011 and 2030 to ensure safety and improve airspace efficiency.

The UK controlled airspace system is an integral part of the European ATM network, where aircraft are ascending and descending into the same small volume of airspace. Development of proposed new EASA rules on airspace operation, management and design, in close cooperation with Single European Sky rules on airspace matters such as the UK/Ireland Functional Airspace Block must be negotiated by the CAA in its trusted partnership role with EASA. EASA has no remit to interface with the Military use of ATM, so the UK CAA and UK NATS must surely lead here.

The UK must use the development of the FAS as a vehicle through which to articulate common strategic challenges and influence how they might be addressed. Regulatory engagement will be essential. To allow for the effective implementation of the proposals in the FAS, the UK needs to press for the right balance between decisions to be made at the European Level with those to be taken Nationally.

The UK Government's Aviation Policy framework refers to none of these considerations at all since it was published in 2013 and has not been updated following the EU Referendum in 2016, but it did support the SES and the UK-Ireland FAB. It also supports ICAO, EASA (the EU's European Aviation Safety Agency which we will need to maintain our support for), and the EU's ETS (Emissions Trading Scheme which we just have to recognise in a world context). Otherwise, I support it on its UK internal market (airport noise etc) considerations seeking increased economic activity.

5) DEFENCE

a) **From:** Air Chief Marshal Sir Jock Stirrup GCB AFC ADC DSc FRAeS FCMI RAF

Ref: D/CDS/4/4/6 Chief of the Defence Staff 19th May 2008
LOGO

To: Mr Jim Quinn BSc FIMechE

Dear Mr Quinn,

Thank you for your letter dated 4 May 2008 and your concern regarding the future availability of aviation fuel. You are correct in stating that the conversion of coal produces a high quality aviation fuel; unfortunately this process also produces significant amounts of greenhouse gases, particularly methane, and is generally considered to be broadly twice as environmentally damaging as the refining of crude oil.

The production of fuel from truly sustainable sources that will demonstrate an overall improvement ('well to wheels' or *well to wing') in their life-cycle carbon emissions is the subject of much current research. The production of fuel from biomass that would otherwise be wasted is one of several attractive options for the future. The use of electricity to power road transport is clearly an option but the need for significant volumes of fuel to be used in the power stations to generate the electricity required must be recognised, as must the environmental impact of this electricity generation. Electrically powered vehicles are also of limited military utility as there is no guarantee of the essential supporting infrastructure being available in those parts of the world where we may find ourselves conducting operations.

Defence will continue to be a significant user of oil derived fuels for the foreseeable future, not just in our aircraft, but also in our ships and land vehicles. However, Defence fully recognises the need to contribute to a more sustainable approach and as a reflection of this, fuel has recently been included as a specific topic to be addressed by the MoD sustainable Procurement Programme Board where both reduced consumption and alternative fuels will be addressed.

Yours sincerely
Jock Stirrup

b) OUR NEXT UK COMBAT AIRCRAFT

We have a Tornado and Typhoon aircraft because MoD invested money in the companies that make it. They are both very capable aircraft and Tornado is still flying in Deutschland, Italia, and Saudi Arabia, probably until 2030. Typhoon carries very few bombs, and nothing like Tornado's 10! The Government gets tax returns from the employees and from the businesses involved. This reduces the cost of a Tornado or Typhoon, and Government then pays a relatively low price for their production aircraft accordingly. Government also said that Tornado prices were just about on target compared with predictions, and they complemented the process involved – British Aerospace et al, and their own people no doubt who were employed by MoD to assess proposals relative to requirements.

Typhoon is a very successful aircraft and engine, developed out of the Tornado and EAP, but digitally, so why did MoD go to the USA for their next aircraft, now called the F35B Lightning? And going with the USA means you can forget technology acquisition and signifiant UK exports because of USA Trade Terms. The reason was that Deutschland did not want a STOVL (Short Take Off Vertical Landing) aircraft so USA was the only route after their enthusiastic reception of Harrier. However, a STOVL aircraft on our Carriers means that they cannot accept French Rafale fighters, so we have limited ourselves operationally in a world situation. However, our F35B's can land and take-off from French

Carriers presumably. We could have had catapults on our two new carriers, which would then have allowed French and US Navy aircraft on board, but the USMC will be able to land on our two new carriers as they still can on the French and US Navy carriers of course. I think it is a shame that catapults were not installed to better support NATO. Just to show how long it took (26 years because of Government delays) to get to the F35B which first entered service in July 2015 with the US Marines, the following P&W visitor badge shows when I started on it:-

So, Tempest is the next – maybe Europe can go it alone again for manpower and technology gaps are threatening otherwise, with a big enough gap (because of reduced workload in the UK) developing over the F35 already. So, I hope MoDUK go with Europe, whose programmes have been so successful. Unfortunately, France always want Design Leadership which is not acceptable to us nor among our other friends.

6) DEMOCRACY

It was the Vietnam war that turned me against the style of democracy that is the USA President. President Nixon had too much power – he regularly lied to the USA population, and he hysterically launched a bombing campaign more than once when he had promised to stop the

War. The Pentagon just did what they were told! President Donald Trump has acted to ban people from certain countries entering the USA, and their Border Control immediately reacted and stopped people without notice – Mr Trump's declaration should have been agreed by Congress and the correct paperwork produced in the first place. In January 2018, the USA Government was shutdown because they could not agree a budget. I note that it was 1789 before the USA had their first President, but the USA war of Independence finished in 1776.

In contrast the UK has had a democracy of the Barons since the time of the Magna Carta 800 years ago, and we have evolved from a long time dictatorial Kingdom which ended in Charles 1st execution in 1649, through an overpowering Presidency in Oliver Cromwell who died in 1658, into the party system with a Prime Minister such as we have today. The House of Commons was separate from the House of Lords since 1341, and we have had a party system since 1680 (Whigs and Tories) - our first Prime Minister Robert Walpole was a Whig, but the system did not evolve further until 1721 when he took office. However, it took until the 1911 Parliament Act to remove the Lords ability to reject Commons legislation – they could debate it, but no longer reject it which they had done previously. This at last was the democratic model I like, which means that policies are democratically agreed within a party, and then the population votes for the party's policies, with a Prime Minister coming from the winning party, and that PM becoming constrained by House of Commons procedure. We have had our troubles too – the 2003 Iraq War for example was agreed by Parliament democratically, but it was based upon false information from PM Tony Blair, and Prime Minister Thatcher sank the Belgrano in 1982 with her directive, even though it was very far from the Falklands and not a threat at all. The European Union has a Parliamentary system, with unbalanced voting rights distributed among the Nations. I have decided after many years of study, that the UK party democracy system, in the end, is the one I prefer, but we need more enterprise.

7) ELECTRICITY

I am concerned about the state of our electrical energy system. Government's Feed in Tariffs (FiT's) have allowed people who generate their own electricity (from Solar PV for example) to have Government support for doing so, even if they do not feed their electricity into the National Grid. The Solar PV windows on the roof generate DC power which is converted into AC by an inverter in the house. This inverter records the kWh (kilowatt hours) continuously and it is supplied to the Government, and that is what the Government's FiT's support, whether the electricity is stored in a battery or supplied to the grid. If it is stored in a battery, the person generating the electricity does not pay the National Grid at all. So their electricity bill is zero. This means that I am paying for my own electricity plus a big proportion of somebody else's through Government taxes. Why have the Government done this to me, and to all my fellow consumers? Clearly, Government wanted to spread the renewables industry as widely as possible, but this sort of support is surely excessive.

I have looked comprehensively at the seven Parts of the Energy Act 2013, with its latest updates, to find that it is very lengthy at 262 pages, and contains mostly the law as it applies to the contracts, regulations and orders, but not at all any hint of Engineering, which it seems is always delegated outside Government (like to InnovateUK or a University). All Acts (not just the Energy Act) always defer to the Secretary of State, who can be seen to be a powerful person for it is his or her responsibility to actually introduce the Act, and to carry out its delegated powers. Royal Assent (given by the House of Lords) means almost nothing for it is the Secretary of State who authorises each and every paragraph or section within the Act. The Secretary of State does not even authorise the whole Act in one go, but has done it in sections on different dates, using SI's (Statutory Instruments) which have to be approved through both Houses of Parliament. There are six of these SI's issued on different dates between 10th February 2014 to 4th May

2016 for the Energy Act 2013, some of which modify the Act itself. The whole scene is very confusing because of these SI's, but also you should note that the Energy Act 2013 modifies other Acts (not just Energy Acts) from previous years, so unless you have the latest Act (of any sort) off the internet you will never work out where you are! The Legislation Team updates these Acts as modifications are agreed – the "latest available" can be found at http://www.legislation.gov.uk, but it takes time to produce the actual latest of course.

The Electricity and Energy Acts have been tinkered with over the years, recently quite often – for example 1989, then 2004, 2008, 2011, 2013, and 2016, and all seem to cover contractual detail only, and certainly nothing of engineering nor scientific interest.

The table below shows what they were for:

Date	Pages	Reason
1989	260	To provide for the appointment of a Director General of Electricity Supply and to abolish the Electricity Consumers Council and the Consultative Council; to dissolve the Electricity Boards and the Electricity Council introduced in the Electricity Act 1947. Etc.
2004	306	To make provision for the development, regulation and encouragement of the use of renewable energy sources; for provision of nuclear decommissioning and radioactive waste; to make further provision in connection with regulation of the gas and electricity industries. Etc.
2008	195	To make provision about payments to small scale generators of low carbon electricity; and payments in respect of the renewable generation of heat; to make provision about the duties of the Gas and Electricity Markets Authority; smart meters generally mentioned too. Etc.

2011	117	About information relating to energy consumption, efficiency and tariffs; for increasing the security of energy supplies; about designations under the Continental Shelf Act 1964; about the decommissioning of nuclear sites and offshore infrastructure; for the use of pipelines for carbon capture and storage; about the powers of the Coal Authority. Etc.
2013	262	To make provision for the setting of a decarbonisation target range; to reform the electricity market to encourage low carbon renewables; to establish the Office for Nuclear Regulation. Etc.
2016	97	To make provision about the Oil and Gas Authority and its functions; to extend Part 1A of the Petroleum Act 1998 to Northern Ireland; to make provision about wind power. Etc.

Do not expect to see any detailed numbers, nor anything to help understand how the "markets" actually work, or even any named organisation (not even the "National Grid") for the Act speaks in general organisational terms only – everything is up to the Secretary of State!

I heard from Lord Tombs at an Institute of Physics meeting a few years ago, that there were no Engineers in Government and that surely is a concern, but all of the Acts that I have read passing through Parliament confirm that. How did Government agree EDF should go ahead with the Hinckley Point C Nuclear Power station – just financially? I think nuclear power is good for the future and I am not worried by the decontamination and storage of waste issue because the modern station does not produce much waste compared with that already hugely stored from nuclear research of years ago. We should have more nuclear, but how did the Government agree on EDF?!

The Government decided that wind turbines were to be supported, for it heavily subsidised them and the wind turbine manufacturers took advantage of it. Offshore wind appears to be less expensive these days for prices have reduced over time, compared with nuclear but wind turbines only last 25 years, while nuclear lasts 60 years according to Mott MacDonald's DECC report of 2010. So wind turbines have to be rebuilt three times as often as nuclear power stations do? Thus their expense compared with nuclear over a 60 year period to produce the same total output. I would like to know whether there was a delegated authority on Offshore wind turbines thus to know whether DECC/ BEIS had any Engineering Authority over the designs.

From: Lord Tombs LOGO 15th February 2007

Dear Mr Quinn,

Thank you for your letter of 10th February about my speech in the House on 8th February. I admire your energy in rising at 6am to watch the debate!

My remark that "the programme will be cash-negative for around 15 years" does not in fact refer to its investment performance, which I am sure will be highly satisfactory, but to the cash requirements of the programme during construction and early years of operation.

The cash flow is of course a separate consideration from return on investment, which continues for the life of the plant.

For most long term industries, projects are financed in part from internal resources but, when those are limited the pressure on cash flow is great. This will be true in the resumed nuclear programme, as it was for Rolls-Royce following privatisation, when we undertook a huge development of new engines. The benefit of that investment is seen when the strength of the enterprise can cope with investments and the associated cash requirements. But the size of investment for a rapidly expanding business does create cash management problems and investors are reluctant to invest on such time scales.

For the nuclear industry today, the Government creams off two thirds of its cash generated and I was at pains to point this out as a serious problem in financing the expansion which is necessary.

I hope that this helps to explain quite a complicated problem and I am grateful for your interest. Let us hope that the nuclear industry weathers the present situation as successfully as Rolls-Royce has done since 1987.

Yours sincerely
TOMBS

I enjoyed his closing remarks in the debate – perceptive, amusing and very educational! See Hansard for "8 Feb 2007" column's 820 and 821.

Lord Tombs told me in that personal letter in 2007, that Government were creaming off two thirds of nuclear cash generated, so no wonder nobody wanted to invest. Maybe that has changed with EDF acceptance? And where did EDF get all their money from? They agreed a price for their delivery of electricity in many years time, and that is where their income will come from, but their bank loans and current earnings must be huge to cover them for say 8 years build time, before future earnings obtain an income. Did all the Government level engineering exist in the Office for Nuclear Regulation? I have discovered that InnovateUK is the delegated authority for support of business R&D, and they do have Engineers to assess proposals for funding. So should we be concerned about the lack of engineers?

Yes, I think Engineers are in delegated authorities, but good Engineers can also think well about many things other than engineering because they are also involved in organisation and money. I need to know how much Government does delegate and what the rules are, and how much it does itself. I consulted the "Planning Our Electricity Supply" White Paper, 2008, and it concerned me that all they were talking about was the market, and it said that the market (which one?) was not working, but they did not propose an alternative to those markets. There was nothing about different technologies in the White Paper, so presumably they do not get involved in those decisions. It has taken EDF years to make a proposal, and success was presumably based on the agreed price for their electricity. How was the price agreed? Were any Engineers involved in assessing the maintenance needs of EDF, and therefore to evaluate the cost to EDF of such activity, let alone repaying the loans they must have had to negotiate, which only an accountant would be able to satisfactorily resolve?

I enjoyed Lord Tombs book ISBN 978-1-84885-506-9 "Power Politics".

8) THE RAILWAYS

In the UK's privatised rail network, a franchise is awarded to the bid which is deemed most viable, and which offers the best value and reliability. Bidders' past performance is also considered. Clearly the franchisees bid against each other before being awarded a franchise, and a judgement is made by the Department for Transport about who would be best. Performance is monitored throughout the contract period of course, so here again visibility of a relatively small organisation would be better than that of a huge nationalised conglomerate, even if it was broken down into regions.

The collapse of a rail franchise under privatisation is a problem, but it would have been concealed under nationalisation. The advantage this visibility gives to politicians is that they can more easily complain about a franchise, than they ever could about the old nationalised regions. Visibility is what counts, not proposing nationalisation because a franchise is in trouble. Those in trouble are visible and thus can be sorted out given the right investigative work. It needs good Engineers,

Accountants and Managers to sort out a failing franchise, and the Department for Transport effectively nationalises that franchise until a replacement can be found.

ISBN 1-85410-823-9 "Broken Rails" paints a picture of a poor strategy by Government for many years, and suggests that Railtrack be re-nationalised, but this is a 2001 view, and stability since then has resulted. I note that Network Rail, which replaced Railtrack, was formed in 2002 and its accounts can be seen in Companies House alongside all of the other franchisees accounts – TOCs (Train Operating Companies) and ROSCOs too (Rolling Stock-leasing Companies). I note, from Companies House, that Network Rail does not pay dividends to its shareholders, and the disaster that was Railtrack, which put shareholder dividends ahead of everything else, has been nullified so nationalisation is no longer necessary, and which in any case would cause another extreme hiatus in the industry.

The Office of Rail and Road, ORR, now regulate the rail industry's health and safety performance, they hold Network Rail and High Speed 1 (HS1) to account, and they try to make sure that the rail industry is competitive and fair. They also monitor Highways England and have economic regulatory functions in relation to railways in Northern Ireland and for the northern half of the Channel Tunnel, situated in the UK. The ORR safety and economic functions are driven by EU and UK legislation. They operate within the framework set by UK and European legislation and are accountable through Parliament and the courts. Their long-term vision for the mainline railway industry is a partnership of Network Rail, operators, suppliers and funders working together to deliver a safe, high performing, efficient and developing railway.

My investigations at http://www.recenttraintimes.co.uk/ show that the current rail services often run just over 5 minutes late but not by much and they are often on time too, and no way could nationalisation beat that. The importance of running on time is now a key factor in a Train Operating Companies thinking as directed by the ORR.

I think HS2 should start in the North and only later build the London to Birmingham section!

CHAPTER 19

PROPOSALS FOR GOVERNMENT

I have thought about my experiences over the last many years and have produced a short list of those policies for Government that I think are worthy of debate:-

ONE

1. Policy is above the Law. The Law was agreed some time ago, and the World has now changed, and new Policy has emerged. Clearly the Law must now change because of the new Policy. Hence Policy is above the Law.

   The Queen should be allowed to retire if she wants to, and Prince Charles should take over, but in my view the Queen should remain as Queen Elizabeth II even while Charles performs the formal role of Monarch as King Charles. This might be a good policy, and it resides above the Law. If the Queen wants to do this, we must change the Law, so that She can decide for herself – but we should volunteer to do this for Her, so that She does not have to ask.

TWO

2. The Highway Code should be mandatory in Law without exception, except for the very special Qualified Police Driver who may exceed it with permission.

THREE

3. Engineers look after their Customers, Lawyers look after the Company against the Customer if necessary. Engineers aim for the bull on the dartboard, as if it were the best moral code, while Laws are built around the outside of the dartboard to limit the criminal's behaviour, and some Lawyers think only in those terms, not necessarily at all in the bull's best moral code. We should object to the legal terms used on software downloads from the internet for they are protecting the Company from everything including their own liability for getting things wrong, and are damaging to the Customer.

FOUR

4. The Prime Minister should surely respond to every letter submitted, either by delegating to a Ministry or by personal team response. If the Prime Minister wants to stop you writing aggressive letters, then a clear stop letter should be sent. The recipient might then argue, but at least knows the Prime Minister's opinion.

   The following chart shows how the organisation should work. It shows that the Prime Minster's website should respond with the email submitted, because otherwise you have no record of what you sent in. I heard that The Prime Minister received one million letters a year (20,000 a week), and I thought he ought to employ enough staff to respond to everyone. I worked out that, if every letter took 15 minutes for his team to respond, that he would only need 120 staff employed with that always respond directive.

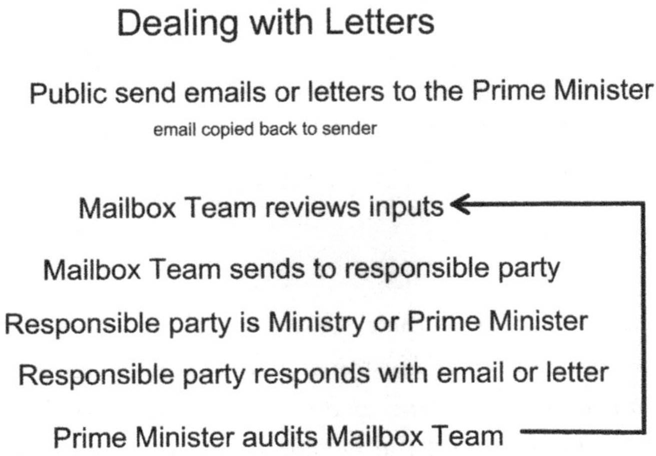

FIVE

5. The House of Lords should surely be filled with genuine Experts, who will mostly be unused to the election process, and therefore they should not be elected, for elections will not be possible without political party support, which leaves crossbenchers and non affiliated Lords and Lady's with no route to the House of Lords. They have a genuine job to do, addressing Quality in everything we do, and they should be encouraged by Government to help with their experienced advice.

   That great man Michael Faraday said in 1831 "*A philosopher should be a man willing to listen to every suggestion, but determined to judge for himself. He should not be biased by appearances, have no favourite hypothesis, be of no school and in doctrine have no master. He should not be a respecter of persons, but of things. Truth should be his primary object. If to these qualities be added industry, he may indeed hope to walk within the veil of the temple of nature.*"

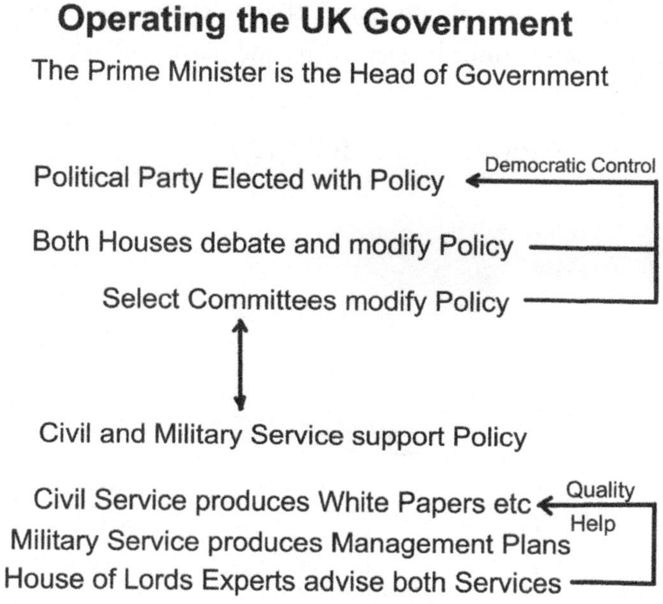

SIX

6. Patents should be free of charge, for the inventor today often cannot afford a Patent application from its initial submission to its years afterwards payments, and business ideas are lost thereby. When Patents are published the information is communicated world wide, and other people in other countries can easily steal the ideas and adopt or patent them in their own country, without reference to the original inventor. So, Rolls Royce plc considered whether to Patent anything at all, for their ideas might go to their world competitors like GE or P&W.

   My 1999 patent below, was far too expensive over the years, several hundred £pounds each year! The bell mouth intake reduces the pressure losses, and the mouthpiece was similarly shaped internally. The small diameter tube was deliberately designed to enable easy evacuation of any ingested water by blowing out hard, and needed the bell mouth to reduce losses into the mouth. This I proved to be quite effective, and much better than the modern large diameter snorkel which cannot be cleared in one blow.

UK Patent GB 2 300 813 B

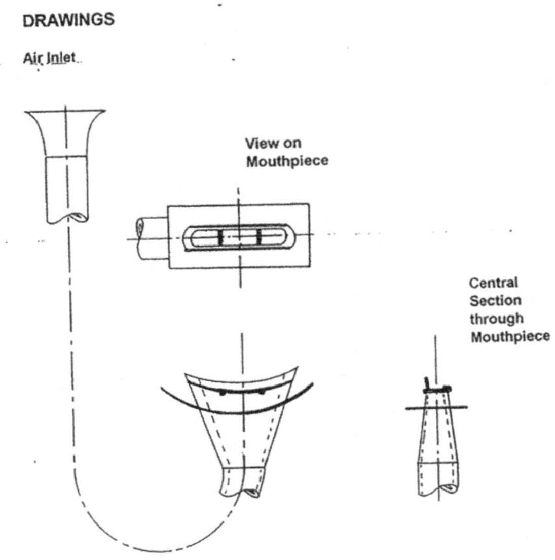

SEVEN

7. The teaching of all kinds of religion including Humanism should be mandatory in all schools whether Christian or Muslim or Hindu or otherwise. The UK Government are trying to introduce a rule saying that faith schools can have 100% of children of that faith (compared with the only just acceptable 50% rule today), so that those kids may never be exposed to other religions. Northern Ireland's problems have mostly been because of such religious stereotyping, and the need to mix children of different parental faith has been recognised for many years now to benefit the community. So whatever are the UK Government doing? Very wrong surely.

EIGHT

8. The Library Service should be supported for it is a local and central source of information which people can use, especially children. Many libraries run school sessions to encourage children to read, and to learn what resource is available. I find it useful to save me buying books, for I tend to read about many different subjects! And often the Library can find books that are no longer available in the seller's market place. Rural Post Offices should also be supported for they have become information centres of interest too.

NINE

9. Personal loans which do not require payment for a year should surely be reduced to at the most just two months, for they cause financial panic after a year when the forgotten loan suddenly has to be paid.

TEN

10. A golden share should surely be considered for important UK companies (as currently exists for Rolls Royce plc), to prohibit majority foreign ownership, because profits of the new company ownership then go overseas.

ELEVEN

11. The best hospitals should surely provide information about how they do it to other nearby hospitals so that the Quality of all is improved.

TWELVE

12. NICE should surely recognise that blindness is the most difficult disability and that treatments (for AMD and WMD for example) are top priority, surely above cancer and all else.

THIRTEEN

13. The House of Commons Sixth Report of the Committee of Public Accounts (Session 1986/87, relating to the Control and Management of Major Equipment) attributes the following statement to Peter Levine on 15th December 1986:

    "If you look at the Tornado, it is an excellent example of how to do these things properly and we expect and hope that the new European Fighter aircraft will follow in its footsteps." Thus I hope the new Tempest fighter is a European one at the concept and development stages.

FOURTEEN

14. Chartered people definitions should surely be refined for Government has reduced its Quality in 2004, by calling some London teachers "Chartered" thus to allow them to be paid more, but without earning their status. The National Union of Teachers said it made money available for teachers in "one part of the country that is not available to teachers elsewhere". We should review the use of the word Chartered for Institutions too.

FIFTEEN

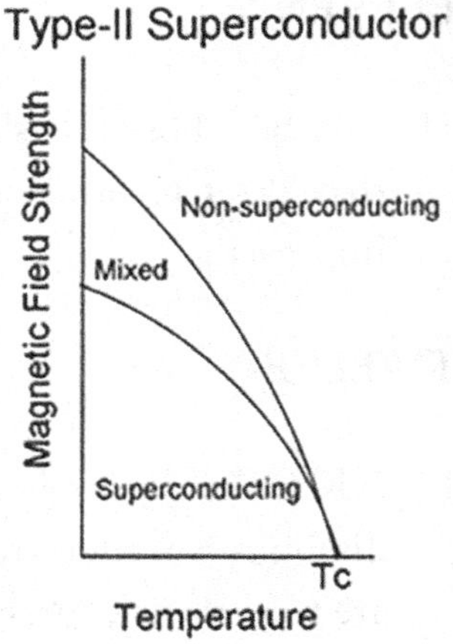

15. Fusion power, as in ITER in Cadarache in south east France and JET in Oxfordshire, should surely be supported through Euratom, and we need to consider limiting the widespread use of Helium for it may be needed in future fusion power stations for superconductor cooling. See ISBN 0-7484-0892-4 "Basic Superfluids". I also think that the Niobium-Tin

superconductor with a Tc of just 20Kelvin with Helium liquid at 2K-3K as the coolant on ITER should be replaced in DEMO (the power plant successor to ITER) by the much higher 90K Tc YBCO ($YBa_2Cu_3O_{7-X}$) superconductor with Nitrogen liquid as the coolant at about 20 Kelvin initially. I imagine the ignition start up of the 500 MW plasma in ITER may cause a sudden large heat flux which will raise the initial temperature of the Helium coolant, and we will need as much magnetic flux from the superconductor as we can get then. If the Helium superconductor coolant starts liquid it could easily turn into gas as the plasma lights, for it will only take about a 1°C degree temperature rise to do so. However, maybe the rate of ignition heat release can be reduced in ITER, but the problem with ITER is that quench of the superconductor may result during ignition. If this occurs, the material will no longer be superconducting, and the electromagnet could be destroyed – a risk with ITER being so close to the limits of superconductivity. The advantage that a high Tc gives is that the magnetic field strength of the superconductor can be much increased, giving more control over the plasma. The difficulty with the structure of YBCO is that it needs more research in manufacturing processes to make it suitable for very large superconducting electromagnets, but it was discovered in 1988 – see ISBN 978-9810232856 "Microstructures and Structural Defects in High-Temperature Superconductors", and things have moved on.

ITER will not produce electricity for it has no generators, and it will therefore only run for a maximum of 6 minutes because it will overheat – what a waste, for we need to control the plasma for much longer than that if fusion is to generate electricity usefully. It is a research scientists dream and not practical enough. So, I am sceptical that ITER is worth all the money that has been ploughed into it! They should have had some broad-minded Design Engineers involved to produce a working model – DEMO is planned for far too late.

The European Atomic Energy Community (Euratom) was founded by the Treaties of Rome of 1957 with the aim of creating a European market for nuclear power, to ensure free movement of capital for investment in nuclear energy and free movement of employment for specialists in the sector. The main need for Euratom is therefore the careful control it has over the use and storage of nuclear fission materials today as well at the specialist workers training and involvement – Brexit is already causing concern over the Government aim to remove the UK from Euratom, which I do not support, even if it does mean that the European Court of Justice, ECJ, continues to govern Euratom legislative adherence, though the ECJ over very many years has never needed to be involved.

SIXTEEN

16. Trolley bus technology should surely be explored for lorries too, for those days when we are all electric and oil becomes very expensive. Hydrogen fuel cell technology needs to aim at minimising cost.

SEVENTEEN

17. The UK Public Inquiry should surely use slides more frequently so that the audience may participate more than the lawyers voice only drone allows today.

EIGHTEEN

18. I have thought about my career and those I have been friendly with, and have written the following:-

    There are too many University graduates with insufficient practical experience, for they appear to have great difficulty in finding a well paid job, and tuition fee payments are triggered at too low a salary level. Some students I know, achieved a good BA (Hons) in European Studies with French, but even though Europe was high on the agenda then, nobody wanted that degree because the job market dropped out at that time. There were no relevant apprenticeships at that time to support continuity of ability into a job.

Apprenticeships support people who want their practical experience to be gained in a thin sandwich (6 months academic, 6 months workplace for 5 years) at the same time as their qualifications are achieved, as mine were (BSc Hons) Upper Second, for this type of apprenticeship provides the best development of your ability, and most often a job will result. A 3 year university career without overall business planning of practical experience interleaved, potentially means relatively poor work experience, and a graduate may not be able to do anything other than plump for a company he or she does not know well when applying for a job. The apprenticeship schemes available today should surely be well supported from craft to student undergraduate levels, and you are paid from day one. No big loans here!

NINETEEN

19. I particularly support the Howard League's proposals to improve the lives of prisoners and their families. I should also say that some of the short sentences handed out to people who rape and murder are wrong, but the Howard League proposals do not alter that opinion.

The criminal justice system is in crisis. Overpopulated Prisons are under immense strain with violence, self-injury and deaths all at record high levels. The partially-privatised probation system is failing to give people the support and supervision they need, and the public is being put at risk. In the 12 months to September 2017, there were 42837 incidents of self injury in prison, 28165 assaults on fellow prisoners including 7828 assaults on prison staff, and 70 prisoners lost their lives through suicide.

a. Community sentences are more effective at reducing re-offending and are one-tenth of the cost of a prison place.

b. Eighty-five per cent of women sentenced to prison last year were convicted of non-violent offences. Women's centres are proven to help women turn their lives around and reduce re-offending, and they need more support.

c. Very few of the children left in the system require custody, and local authority units have a track record of success. We should aim to end the use of prison, and ensure local authority units for children are well supported.

d. Between 2010 and 2015, over a million additional days of imprisonment have been imposed for minor infringement of prison rules - 84 per cent do not involve violent behaviour. These unfair additional days should be stopped.

e. Earning release through parole is crucial for many prisoners, yet the Parole Board does not have enough resources (needs more) and as a result has a long backlog of cases, delaying release.

f. The increased use of recall to custody for very minor rule breaking after release, is one of the main drivers of our ever-growing prison population and overcrowding problem. We must stop this.

I do not understand why the UK Government is so slow to adopt these proposals for they are both reasonable and good.

TWENTY

20. We have been subject to austerity by the UK Government now for just over ten years and unfortunately I think this has resulted in midwives leaving their profession for lack of pay, finding jobs elsewhere. The situation has become chronic in that now midwives are leaving because they are over worked as a result of lack of colleagues. This will put mothers in danger, because they will not receive the caring they should expect and some will die. This will leave the midwives who were trying to care for that mother distraught and fearful they will be prosecuted, but I argue that it is the lack of attention from the UK Government and NHS Executive that has led to this crisis. There are more midwives leaving the profession than are joining it. How do we recruit and train more midwives in zero time?

    I think we have to divert NHS staff to midwifery – I choose NHS staff because they are medically trained already and should only need a short training period to get up to speed. This has to be a priority above other areas who suffer less than the midwives do.

    Whilst I acknowledge that the UK Government were concerned about finances and introduced austerity, there has to be a balance struck between possible deaths of mothers through lack of treatment and investing in other programmes, like education. Teaching staff are not in the same life threatening situation that midwives are. Therefore I think additional funding of midwives should have started years ago, supported by Education Department Policy.

    Austerity itself comes about from a lack of GDP – somehow we need to increase our GDP and in the short term I think we have to borrow from the IMF to fund investment into business and thus gain exports for the longer term to repay those IMF loans.

TWENTYONE

21. Have you ever wondered why pedestrian traffic lights make you wait in the wet and cold while warm and snug car drivers keep going through the lights until, after a long delay the lights let you cross? Yes, typical Civil Service, not thinking that the delay should be <u>after</u> the lights let you cross not before it. My way, if you are the first to arrive after a time, the lights will immediately change for you the pedestrian, instead of callously ignoring him or her, on a dark night when all sorts of threats may surface. And it would save so many car drivers becoming frustrated and angry waiting at red lights when nobody crosses, since the pedestrians have already done so between cars (at risk) having been bored waiting.

TWENTYTWO

22. This is not a Government proposal, but a TV proposal – I become very frustrated at not knowing who the person being interviewed is – the TV people show the name so briefly that I cannot read it – I need it for the whole interview please, then I can read it and know who it is when I come in to watch the TV having been out of the room for a while.

TWENTYTHREE

23. This is also a TV proposal, for I become frustrated when TV interviewers do not let their subjects continue with a sentence about their thinking, and are interrupted even before they have managed to express their reply to the previous question!

TWENTYFOUR

24. This is also a TV proposal for I become frustrated in programmes like the Shetland detective drama, that the camera does not stay on the subject when being interviewed and I find it very difficult

to lip read when they so rapidly switch back and forth between the interviewed and the interviewer. The BBC seems to have a particular difficulty in understanding what communication is all about.

CHAPTER 20

A UK CONSTITUTION

A new Constitution has been proposed below, because nobody, not even every lawyer nor every MP (who have to refer to the Attorney General to find out about it), knows the present UK one (rumoured, for I know nothing of it having tried to find out about it). Because there is no Constitution that is available to all, immigrants (and Muslims with their women and heirs restrictive Sharia Councils) have no idea what we are about either. There are also about one million illegal immigrants in the UK, and none of them have sworn the (very simple single sentence imparting very little policy) Oath, Pledge or Affirmation that legal immigrants are required to do – taken from Schedule 5 of the British Nationality Act 1981 as amended.

Even legal immigrants therefore do not know our Constitution, alongside the rest of the 65 million population of the UK, who do not know it either, and have not so sworn. This surely emphasises the need for some good Constitutional words written down so that they can be understood by everybody, and thus to know what our UK principles really are. I propose my draft of words in THE LOGIC below, for they are simple enough to understand and do not occupy too many words, so can be generally remembered.

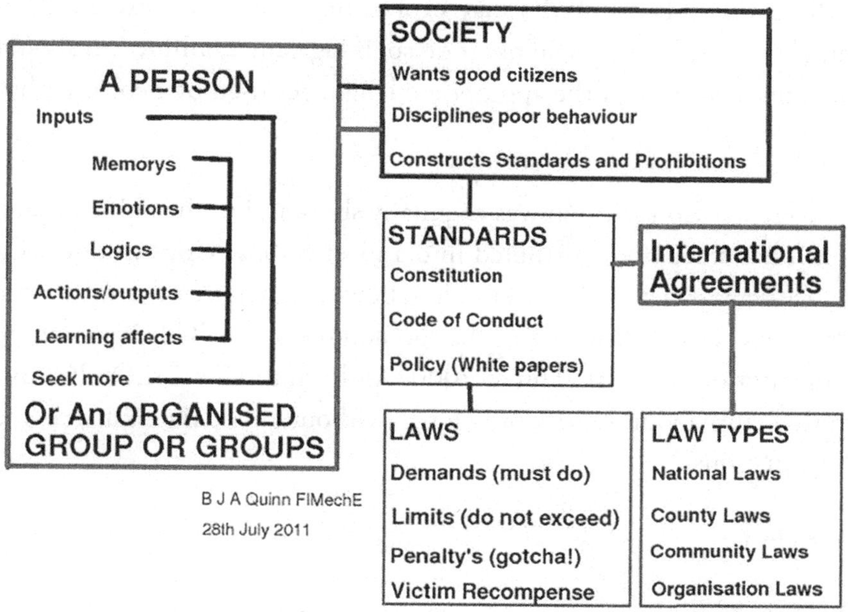

The Proposed Constitution

THE LOGIC – Introduction

I believe in individual rights and freedoms, but that individual responsibility, social cooperation and mutual respect are also important.

I have an approach to life based on humanity, real data, and reason – I recognise that moral values are properly founded on human nature and experience alone, and are based on the available and sought for evidence, and our assessment of the outcomes of our actions only.

I believe that people can and will continue to find solutions to the world's problems so that quality of life can be improved for everyone.

I am positive, gaining inspiration from our lives, art and culture, and a rich natural world.

I think that THE LOGIC may evolve, but must never exceed 2000 simple words. It surely will never do so, if logically examined. It should be translated into all the appropriate languages used by people within the UK.

The French Constitution was originally short and to the point as here, but it has since been corrupted into a most verbose Constitution with non essential detail. For example, about the length of time after a President is elected that he or she should take office. This is surely not Constitution material. Thus a 2000 word limit to THE LOGIC, this draft being about 1430 words long (without this paragraph nor the Explanations).

Article 1

THE LOGIC is the expression of the general will, here enacted. All citizens have the right to personally take part in its evolution through a referendum. It must be the same for all, whether it protects or punishes.

Article 2

Men and women are born, facially recognisable, and remain free and equal in rights. There are too many people in the world, and a plan for the whole of humanity must be evolved, so we must stop having so many children, and we must stop the spread of genetic disease. Homo Sapiens is to practice the elimination of genetic diseases by careful selection of embryo when circumstances permit.

Article 3

The Prime Minister is the Head of Government, and will be a member of the House of Commons. The Head of State will only do what he or she is allowed to do by the Head of Government, and takes no part or action in political thinking. Her descendants may be political if they wish.

The House of Commons will be entirely composed of democratically elected People, who will voice their feelings upon any Act, Bill, Plan, Rule or Procedure in debate, and who will ensure that the Policies are democratically acceptable.

There will be a House of Lords comprised of Experts in many disciplines who are not elected, but will have been nominated for their ability. The House of Lords will use its expertise to introduce, examine and amend all Bills, Plans, Rules or Procedures passed by the House of Commons, using their expertise to do so, if necessary passing them back to the House of Commons for further debate. The experts in the House of Lords will also advise the Public Services, that is the Civil and Military Services, to ensure the highest Quality and that all logical potential solutions have been addressed.

All citizens shall be equally eligible for all high offices, public positions and employments, according to their ability, and without distinction other than that of their character.

Article 4

The City of London and all other major Cities shall have elected Mayors to act as Leaders and Managers in their own right within Central Government guidelines. (Explanation: This Article is adapted from the Magna Carta).

Article 5

The aim of every political association will include the preservation of the natural rights of every man, woman, or child, as here enacted in THE LOGIC. (Explanation: Thus there will be no Nazi party nor Daesh).

Article 6

The source of all sovereignty lies essentially in the State. No corporate body, and no individual, may exercise any authority that does not expressly emanate from it. All citizens have the right, through their elected representatives, to decide the need for and approve simply written Law, which is subject to THE LOGIC.

Article 7

Liberty consists in being able to do anything that does not harm others. Thus, the exercise of the natural rights of every man or woman has no bounds other than those that ensure to the other members of society the enjoyment of those same rights.

Article 8

Everyone has the right to freedom of assembly and association, within the bounds of THE LOGIC.

Article 9

The Law must prescribe only the punishments that are strictly and evidently necessary, and execution is prohibited. No one may be punished or exiled except by virtue of a Law drawn up and promulgated before the offence is committed, and legally applied. Those who solicit, expedite, carry out, or cause to be carried out arbitrary orders must be punished, but not necessarily by imprisonment. Any citizen summoned or apprehended by virtue of THE LOGIC, must give instant obedience, since resistance makes him guilty. They must be honest.

Article 10

As every man or woman is presumed innocent until he has been declared guilty, if it should be considered necessary to arrest him, any harshness that is required to secure his person must be severely curbed, including accusation without proof.

Article 11

No one may be disturbed on account of his opinions, as long as the manifestation of such opinions does not reject THE LOGIC, which welcomes helpful clarification.

Article 12

The free communication of ideas is one of the most precious rights of man or woman. Any citizen may therefore speak, write and publish freely, except what is tantamount to the abuse of this liberty in the cases determined by The Law.

Article 13

To guarantee the Rights of the Citizen in Law, Public Services are necessary, but they may delegate to private business if appropriate. These Services are established for the benefit of all, and not for the particular use of those to whom they are entrusted. (Explanation: thus corruption is not allowed.)

Article 14

For the maintenance of the Public Service, for Public Investment and Personal Benefits and Pensions, a general tax is essential.

Article 15

All citizens have the right to decide through their elected representatives the need for simple public taxes, to consent to pay freely, to watch over its use, and to determine its proportion, basis, collection and duration.

It must be collected from all citizens and businesses, in proportion to their income and to their expenditure.

Article 16

A Citizen has the right to ask a Public Official for an accounting of his or her administration.

Article 17

Since the right to property is necessary and traditional, no one may be deprived thereof, unless public necessity, legally ascertained, logically requires it, and just and prior indemnity has been paid.

Article 18

Everyone has the right to health and welfare care, as determined by the State and the individual alone, and the adult may decide their own future for themselves, including the right to choose when to die peacefully.

Article 19

Everyone has the right to the protection of personal data concerning him or her.

Article 20

Everyone has the right to a personal education to bring out their natural ability and talent and upon which that personal education is based, and for that education to continue throughout Life, as determined by the State and the individual alone.

Article 21

The use of the Earth's Resources, whether of climate, organic or inorganic materials, or people, must be carefully husbanded, to provide a measure of sustainable fun for the Future.

Article 22

War is the most expensive action of last resort to fight the bully. All our weapons must conform to our International Agreements on the type and use of such weapons. War can devastate both ourselves and the opposing Nation or Terrorist. Sanctions must be targeted at the leaders of such a Nation, not more generally on the Nation, because that would hurt the downtrodden person as much as the Leaders. We will always prefer to Trade with other Nations even if they violate THE LOGIC's principles, for we have the opportunity to continue talking about peace, and benefiting the downtrodden population of that other Nation through the extra employment that Trade offers. Our Trade in weapons will also conform to International Agreements on the types of weapons sold to other Nations. We must encourage other Nations to adopt our principles.

Article 23

A child is a person who needs adult protection, love and nurturing. A child is here defined as a person under the age of 18, who may be employed from the age of 13 with a maximum of 12 hours a week, gradually increasing until at the age of 16 and over the child may be in employment for a maximum of 40 hours per week. Those persons over 18 years old will be entitled to vote in Government Political Elections. A licence will be required for children under the age of 13 to be employed, from which age they will be held responsible for their own actions. The threshold age of criminal responsibility will be 13 years.

Article 24

The Orang Utan is our 15 million year heritage, for we have evolved from them. It and other extremely rare animals must be protected from extinction, their natural environment protected, and their rights, as predecessors, the same as here enacted in THE LOGIC.

The famous A.C Grayling in his book The Good State, On the Principles of Democracy ISBN 978-1-78607-718-9 would also like a new Constitution, but for different reasons and with several new Articles.

CHAPTER 21

NUCLEAR WEAPONS

Following Mr Putin's war in Ukraine I have been considering UK nuclear weapons Policy. If we did not have nuclear weapons, I think Putin would have nuked Kyiv by now.

He is clearly a bully having increased the status of his nuclear weapons forces, just at the time he invaded Ukraine. I am glad we are supplying weapons to Ukraine to help them defend themselves, but they need more from us – we must increase our Defence Spending to do this. We must also retain our Nuclear Weapons to hold the bully back from his endeavours. I note that UK Policy is NOT to make a first strike against Putin, but to respond aggressively, and I support this. I am also concerned about MAD (Mutually Assured Destruction) and I see no reason to increase the numbers of our nuclear weapons.

CHAPTER 22

EVOLUTION AND ET

Evolution is a FACT. It applies to the Universe since the Big Bang, it applies to our Sun 8 billion years later, and to us 4.5 billion years after that. And no wonder the Ancients used to adore the Sun, and adopted it as their God, for it gave life to everything on this planet Earth. The last common ancestor of monkeys and apes lived about 25 million years ago, while the last common ancestor of humans and chimpanzees lived about 7 million years ago.

Evolution is the result of random biological change (mutation) constrained by the more powerful Environment.

Random biological mutation can be seen in all species, but it may help to visualise it by looking at the Big Cats – the Leopard, the Tiger, the Lion, and the Cheetah. All of these were one species millions of years ago, but their coats are now all slightly different. This is the mutated safe biological change I mean. Other biological changes within the Big Cats over very many years, caused them to be sufficiently different so as to not mate with each other, and they became different species. In similar fashion, Homo sapiens is the latest part of the Great Apes genre, but the Orangutan, Gorilla, and Chimpanzee all continue to survive too, and the Orangutan have been here for about 15 million years. All because of random biological mutation. By "random" I mean changes that have occurred without planning of any sort – that just happen. These species all survive today because those random biological changes did not cause them to die early, but if any did die before giving rise to their own offspring, that biological change would not have survived. Random biological change can cause early death as well as continued modified survival. The term "survival of the fittest" is definitely wrong for otherwise there would be no species left on Earth other than the insects!

L H Hawkins in ISBN 0-85498-521-2 "Electric Field Phenomena in Biological Systems" suggested that geomagnetic pole reversals occurring throughout the last 65 million (Cenozoic) years, caused rapid biological mutation and thus evolution of so many species since the dinosaurs. DNA reproduction errors can easily lead to significantly different offspring, but this is different from sexual reproduction which merely mixes parental genes without reproduction errors. Note that the north pole going to the south or back to the north is defined as a reversal, but both took many years to change, with poles in all sorts of orientations on the way to north or south. I have not found any evidence about whether there is a pulse effect of the geomagnetic field during these reversals, but some reversals are accompanied by very low field strength (<5% of normal). Evidence suggests that low level pulsed magnetic fields can adversely affect development of chick embryo, and zero magnetic field strength can result in a depressed effect on insect reproduction rate and animal metabolic activity (it resulted in early death in some animals) and NASA in 1966 measured the depressing effect on human central nervous systems. The only other genetic evidence I can find of a measured geomagnetism effect is in the work by Dubrov, who investigated the Earth's local variation of geomagnetism on the fruit fly Drosophila, and found that their gene sequences (measured from the salivary gland) were thereby affected. Demonstrating mutation is feasible with insects because they reproduce rapidly, but it is difficult to extrapolate this to animals of course. Wikipedia says recent geomagnetic reversals occurred 5.25, 4.17, 3.59, 2.59, 2.00, 1.78, 1.06, 0.78, and very briefly 0.41 Mya (Million years ago), but the fossil record is not as accurate, so statistical correlations with mutation or extinction are questionable.

Selected low frequency and high voltage electromagnetic fields can affect animal feelings (including loss of libido, and pain) and lassitude – these are EMI caused behaviour changes. Tests at up to 10GHz produce big increases in neuron conductance, which leads me to think that hyperactivity, loss of coordination, and hallucinations might be so caused, but so could many other behaviours for the brain is very complex molecular nanotechnology, and you just have to trigger the right external frequency to cause a particular feeling or behaviour. Further, the book also covers the logic about how some electromagnetic fields produce faster healing powers, since Ampere's Force can move proteins blocking cell membranes. Then there are 1.5 Tesla MRI scans which align cells in the magnetic field, while Electro Convulsive Therapy used by psychiatrists obviously causes major shock to the whole brain and its neurons. Sharks electroreception is very sensitive to electrical fields emitted by fish - and cameras! And we are now investigating brain electrics with skull caps to better control a prosthetic leg over level and stepped terrain real time and not previously programmed in. Very high frequency signals will not penetrate the skin, and I think only signals below about 1GHz will fully penetrate the brain. All of these things are affected or caused by electromagnetic fields because all brains are electrical in character - in humans the brain consumes perhaps 40W of energy, generating 1 picoTesla (one millionth of a microTesla) of magnetic activity too.

Techniques for electrical recording of neural activity and brain stimulation were just starting to be developed in the 1870s – about 150 years now, but evolutionary knowledge of the brain has needed the latest 20 years of techniques in microscopy, molecular analysis, phylogenetic trees, and gene sequencing to help evaluate a broad range of vertebrates and invertebrates. ISBN 978-0-309-26175-3 "In the Light of Evolution, Volume VI, Brain and Behaviour" says that voltage dependent potassium ion channels appeared 3 billion years ago in bacteria. The vertebrates originated about 525 million years ago (525Mya) during the Cambrian explosion, but neurons with sodium ion channels are known to exist from over 600Mya in eumetazoans, which have functional nervous systems. Primitive tripartite brains in bilateria date from about 520Mya, and similar brains have evolved numerous times independently – thus ET should certainly have evolved one. Myelin wrapped neurons evolved about 450 Mya, and so must have thus allowed bigger and faster brains. By about 200 Mya the typical mammalian ear had evolved, but sounds could be heard by about 350Mya when animals first moved onto land, although not to the higher pitch that we are able to perceive which evolved about 40 Mya.

ISBN 978-0-552-77807-7 "Life on the Edge" is a fascinating and interesting exposé of the world of Quantum Mechanics as it applies to biology, and the book is subtitled "The Coming of Age of Quantum Biology". It uses words like cryptochrome radical pairs and quantum entanglement to explain it, and says that The European Robin is very sensitive to geomagnetic inclination – which is the angle the Earth's magnetic field makes with the earth's surface from zero at the equator to 90 degrees at the poles (25-65 microTesla magnetic field strength, strongest at the poles). All I want to do is to remark how sensitive biology is to the geomagnetic field, and thus possibly to support the geomagnetic mutation of so many species since the age of the dinosaurs at least, but not necessarily every time. It is now thought possible that Pigeons use their sense of smell to locate home, rather than geomagnetics. The effect can be disrupted by high frequency electrical fields.

To explain Quantum Mechanics in a little more detail, bear with me for a moment. Imagine a balloon filled with gas under pressure – all the molecules within that balloon are vigorously moving and banging into each other and into the balloon itself, and if you were to release the neck, the gas would escape with a big whistling sound and the balloon would fly all over the room. Now make a metal sphere the size of the balloon contain just a few atoms of gas as in a substantial vacuum, and put a small atom sized hole in it. The gas will take a long time to escape for the vigorously moving atoms bouncing around the inside of the balloon will not be moving in the direction of the hole, and will take a long time to find it, so it will be incredibly quiet for a time. This is the world of atom (particle) sized Classical Mechanics, and atoms can be seen under the electron microscope.

Quantum Mechanics is more difficult to understand, and conceptually thinks about the nano-small, for it describes the movement of protons and electrons as waves and spins in all directions which cannot be seen inside atoms. The wave effect can be visualised (and has been shown many times in the laboratory) by putting two, very closely spaced, small holes in that sphere filled with rapidly moving electrons at room temperature. When one electron approaches a hole it actually behaves like a wave and passes through both holes at once! The electron is acting as a wave for it can produce an interference pattern on a screen and which can be seen in the laboratory – much as you might, in a particularly shaped water harbour, see where you get wave peaks and troughs amplifying or cancelling each other because they are superimposed on each other.

Quantum Mechanics describes the wave and spin pattern all of the time – except when you observe it to see particles. If you do not look, particles are not particles at all, but waves. Thus an electron is never where you expect it, and you cannot observe any shell that it might be in - protons are thought to be moving waves and spins too, and it is these that affect our DNA, for the normal nucleobases (A with T and G with C - Adenine and Thymine, Guanine and Cytosine) have hydrogen bonds securing them together inside the DNA molecule. A hydrogen atom is merely a proton and an electron, and electrons are already stored on the four DNA nucleobases, so the wave and spin movement of a proton to the partner nucleobase, effectively means that the hydrogen atom has moved – this causes genetic failure in rare cases. The trouble with this is that the bases then do not line up normally with each other, and A in my example has changed so that C incorrectly bonds with it instead of the normal T. DNA replication then fails to be 100% accurate as is necessary for good heredity. This is why we have to be careful with our problem young before they have children, for the whole population may suffer Huntingdon's disease in centuries to come, if we do not perform embryo selection in the laboratory to produce Huntingdon's free youngsters - this may yet be possible. Huntingdon's has resulted from this incorrect bonding.

Quantum Mechanics might also explain the start of biological life on Earth, and I think might be able to do far more than Classical Mechanics could in arranging atoms circumstantially together 4 billion years ago. It is proton wave and spin which never stays still, and I think it can effectively try far more options far more quickly than Classical Mechanics ever could with atoms. Note that I am talking about the first vesicle and the construction of DNA to form the first cell. DNA is formed from the four nucleobases which all contain the atoms from nitrogen gas and (carbon from) carbon dioxide gas in plenty, hydrogen and oxygen from water, and the solid atom phosphorus probably in a volcanic mud-pool, and those gases and water would have been freely circulating back then.

Telepathy is not possible among humans for we all have very similar brains so will generate very similar frequencies which would be required for communication. The mobile telephone network relies on different frequencies to pick out one another's phone, and human brains cannot do that. Sharks are not telepathic, they just electrically detect that a fish is nearby, but do not communicate with it. If we could operate telepathically it would be bedlam for we would "hear" everybody around us all at once. ET cannot be telepathic either for the same reason, but like us they could use mobile phones or similar means of manufactured, transceiver focussed, communication.

The Environment is the really big factor in Evolution. Professor Richard Dawkins once told me in a lecture he gave years ago, about the Peppered Moth. This is a species which produces both black and silver moths, so that if they land on a tree like the Silver Birch, the predatory birds will attack and eat the black ones for they would be so visible, but those birds will not see the silver ones. In similar fashion if the woodland is of dark Oak trees say, the ones to be eaten by the birds will be the silver ones. Eventually, if the two groups of trees are a long distance away from each other, the two colours will become different species, just like the Big Cats did, but note that this takes millions of years to allow more of those random biological mutations to occur – it took 4 million years to evolve the first Gorilla from the time of the first Orangutan. And so on for every species on this planet, as Charles Darwin discovered in the mid 1800's, and documented so well.

Evolution eliminates failed efforts. A monkey that hallucinates a branch where there is none, will soon become a dead monkey. A blind and deaf person crossing a road without a helpful hand on their arm will perhaps soon die. So, in the long run as young death stops offspring being born, an individual with a tragically mutated brain will be eliminated by such natural selection. Darwinian recognition of emotion in apes, has led to an improved view of the use of animals in experiments. We now have good rules for how animals should be treated and many organisations defend the rights of animals. This is Evolution in action too.

I asked Chester Zoo if Orangutans like music, and their Education & Research Officer responded with a 2005 report ("A note on the effect of auditory stimulation on the behaviour and welfare of zoo-housed gorillas" by Deborah Wells, Dwyer Coleman, and Mark Challis). Music has long been employed as a form of therapy for humans housed in institutions, and its effect on one of our closest-living relatives, the gorilla, was explored by determining the effect on the behaviour and welfare of six gorillas housed in Belfast Zoo.

The gorillas appeared to be more relaxed with classical music than with the rainforest (birds twittering etc) sounds. I prefer the wave harmony of music to the non variety of boring rainforest sounds, and I think evolution has significantly made me very much like the great apes, so it is not surprising to me that gorillas also like music, since Homo sapiens likes it so very much.

Logical thinking does not require speech, as many tests on great apes, birds and other animals have shown. Homo erectus probably only had a very simple language – communicating mostly by miming it is thought, but which gave them an advantage in relation to earlier hominids. See ISBN 978-0-19-852851-7, "How Homo became sapiens – on the evolution of thinking". Homo erectus over a million years ago, may also have used fire to cook food - ref BBC4 television "Prehistoric Autopsy".

ISBN 0-521-47168-0 "Reaching into Thought – the Minds of the Great Apes" is an excellent expose of where evolutionists and psychologists have got to in determining how the Great Apes think (these, in evolutionary order, are the Orangutan, Gorilla, and Chimpanzee and Bonobo). Current views lean toward the conclusion that the real divide was when the monkey evolved into the Great Ape, not between the Great Ape and Homo sapiens. Thinking about the Great Ape has only really evolved since the 1960's, and significantly comparisons now recognise that the Great Ape takes more than 7 years to reach adult maturity. The other factor involved is that small numbers of samples could give some negative results – Homo sapiens, similarly, has a considerable range of intelligence from high to low, needing large sampling to be correlated and only then fully representative of the average not the high or the low. Great Apes in more enriched, free ranging settings, also appear to imitate more readily than those in captivity. I recognise the Orangutan as my attractive 15Mya ancestor, and mentally sub-title a good photo as "Grandad", for they have an intelligent look and often a wry smile!

Evolution is well understood by Engineers when they study it, for we evolve our ideas as time goes by, depending upon the results of our experiments and others reported knowledge in published papers too – our evolution is not as random as in biology, as you will appreciate! So, what do I predict for the "Extra Terrestrial" person?

It seems to me that the Environment is what has ensured we have two arms and two legs, with two eyes and two ears in one head. Our primate ancestors had the same, and so did the dinosaurs, and we know whales and dolphins have two arms and two legs, let alone two eyes and two ears. Tetrapods (Sarcopterygii or lobe-finned fish) first evolved about 419 Mya, and it is these which gave rise to two arms and two legs later. And life thus had plenty of time to evolve a different configuration if that was not optimum for larger animals, but it did not. Animals with 6 or 8 legs never evolved into a species which contained a large brain, and they had plenty of time to do that too. We have one head on top of the body so that we can see easily, just as in all other animals. We all have one tongue, one mouth and one jaw too. And gravity has forced the denser intestines to be lower than the lungs in us primates, with the waste coming out in the groin. But gravity might also make the average ET a shorter person than us if he or she comes from a bigger planet – or the other way around for a smaller planet – but Homo sapiens has a big range of heights already! And these things must mostly have evolved initially in fish before they climbed out of the seas to form us land based animals. The eyes – there have been 40 different starts and thus 40 types of eye in the animal kingdom, and all have matured into two eyes in one head (but not in insects). So these configurations seem to me to form the essential shape of ET too. He or she might be a little hairier of course!

And as for the nucleobases A-T and G-C (Adenine and Thymine, Guanine and Cytosine) that make all of our plant and animal kingdom's DNA, that I see as the only solution for ET too. It took a billion years to produce them on our Earth, and it is therefore unlikely to be any other sort of chemical composition that could have been used, nor used on any other planet – silicon was/is as plentiful on Earth as carbon, and silicon compounds are also dissolved in our oceans, so silicon based proteins had plenty of opportunity to evolve - but silicon based species did not evolve after the common ancestor of animals and acanthoecid choanoflagellates about 600Mya – so carbon based proteins are obviously necessary for all animals. We will confirm this one day by visiting Europa and Enceladus very closely, but I am convinced already – carbon based proteins and amino acids A-T and G-C are essential for ET's DNA I am sure!

No, ET will be like us, with a reasoning brain that can think about itself biologically, medically, and psychologically, and about the outside Universe too. To communicate across the vast distances of space requires our intellectual ability, but I doubt that they will be more intelligent than us, for I doubt we will ever evolve a higher intelligence than we already have, with such variation already in Homo sapiens from low to very high IQ, and low to high practical ability too! I think ET would have to have dexterous fingers like us too – to be able to spin thread, to sew and knit clothes, to create finely detailed structures, like screws and nuts and bolts, and the wiring looms of telephone exchanges and aircraft, and radios and radar, and to use the small keyboards of computers, all of which would have been needed to develop ways of exploring space and the deep oceans.

ET will also have our range of skin colour, for the same reasons as we do, but do not imagine they are here already, for the distances across space are so vast that many lifetimes would be needed to cross it from the nearest stars, let alone the likely populated ones, so very much further away. The nearest star is 4.5 light years away, which is about 30 trillion miles, and the Milky Way Galaxy is about 100,000 light years diameter. Our fastest spacecraft have travelled at 50,000 mph, so even if we went twenty times that speed to an unlikely one million mph, we would still take over 3000 years to complete the journey to the nearest star! You would need a spacecraft the size of a big city (with an artificial gravity which will need the whole city to rotate) to live your life on it comfortably there, with all sorts of entertainment, let alone work. And if the nearest populated planet were to be, optimistically, 500 light years away (implying, say, 10,000 intelligently populated planets in the Milky Way), ET would take 3 million years to reach us at one million mph! No, ET will never come here, nor we there. Even at 10% of the speed of light, they would take 5,000 years to reach us! And communications? 500 years each way, so no possibility there either.

Artificial Intelligence is the only way forward for such long journeys, but even that could fail to survive 3 million years, and what would be the point anyway? Artificial Intelligence could survive on Mars and on the Moons of the outer solar system – and enjoy life there doing what humans could not do in less than a thousand years from now! I am not imagining a robot like those already working on Mars, but an intelligent thinking machine enjoying life much as we do. Curiosity is a trait shown in all animals as they search for food, and so artificial intelligence should be easily able to ask questions: but "why is that" is much more difficult to answer than "what is that", and the difficult part is learning to understand and reason with the answer! That is what will take a good many years of research into Artificial Intelligence yet, and "why" is what humans find difficult too.

One other thing to bear in mind is the possibility of hitting small meteorites at 1,000,000 mph (this is 447 km per second). The damage to the ISS (International Space Station) at just 17,500 mph (8 km/s) was already visible in 2013 in a 6 mm hole in one of the solar panels caused by an estimated 2 mm diameter micrometeorite – imagine what that hole would have been like travelling at 1,000,000 mph (= 447 km/s or 0.15% of the speed of light), let alone the thinking of some scientists (including Stephen Hawking) who want to travel at 10% of the speed of light (this is 67,000,000 mph or 30,000 km/s) to get to the nearest star in just 50 years travel time! And with no possibility of slowing down to orbit that nearest star either. Clearly however we should go and explore, so we have to ask the Design Engineers to strengthen the notch capability of the leading edge material.

SPECIAL RELATIVITY

While talking about high speed transit to the stars, I thought you might be interested in this review of Einstein's Paper in 1905 about Special Relativity, for which he did not receive the Nobel Prize.

Einstein produced a Paper in 1905 called "On the Electrodynamics of Moving Bodies", and it is recorded in ISBN 0-691-12228-8 "Einstein's Miraculous Year". It is a logical paper, but the many explanations at the very beginning of the paper on pages 1 to 8, could be improved upon, and are not, in reminder, simply stated individually on page 10 where it would have been helpful so to do – you have to work out what early page paragraph he is implying when you get to page 10. However, I have worked it out as follows:

The first equation on page 10 of the Paper is introduced as follows:

“Suppose that at time $Ћ_0$ a light ray is sent from the origin of the system *k* along the X axis to x’ and reflected from there toward the origin at time $Ћ_1$ arriving there at time $Ћ_2$ – we then must have

$$\tfrac{1}{2}(Ћ_0 + Ћ_2) = Ћ_1$$

”

The equation is obvious for a system at rest, but not so for a system moving at immense speed, where you would imagine that it would take longer to get to Ћ 1 because x’ is receding from you, while it would take less time to get to Ћ 2 because the light is then travelling toward you. If you are travelling at near the Speed of Light, the Light (according to the equation) is being reflected and thus going in opposite directions at the <u>same</u> measured (perceived) speed, which seems impossible, and thus I thought that the equation was wrong.

However, on page 2, Einstein’s paper says:

“Examples of this sort, together with the unsuccessful attempts to detect a motion of the Earth relative to the “light medium” lead to the conjecture that not only the phenomena of mechanics but also those of electrodynamics have no properties that correspond to the concept of absolute rest.”

Einstein is presumably talking about the Michelson-Morley measurement of the speed of light (why did he not say so more clearly?) where no difference could be detected from one side of the Earth’s orbit around the Sun at 108,000 km/hr compared with the opposite direction other side 216,000 km/hr (0.02% of the speed of light) different - his page 2 “unsuccessful attempts”. Note that Einstein extrapolated a long way above that 0.02%, to presume implications at much higher proportions of the speed of light! Thus the speed of light was thought to be a constant by Michelson-Morley, but Einstein actually implies in the equation above, that it is the perception of the speed of light which is constant, not that the limiting speed of light itself is a constant. The speed of light is 300 million metres/second and is what we perceive from our measurements, just as ET on another planet would also do from their perceptions. Thus the light ray travelling to x’ in the direction you are very rapidly moving, will actually only be measured by an observer moving with the rod as the same speed when the light ray is reflected from x’. The equation is a fundamental presumption to the Theory, which would not exist otherwise. This is the Einstein claim of the observer’s perception based upon measurements of time and distance along the X axis. This implies that time and distance are also different in a moving system from that in the system at rest, and is Einstein’s perceived confirmation of Michelson-Morley. It is this perception of his that presumably took so many physicists so long to accept, perhaps because he should have said it more clearly alongside the equation above to help explain the presumption, that their slow speed measurements in 1887 by Michelson-Morley, actually also applied at immense speed. The Paper goes onto show the equations for time and dimension in terms of the speed of light and the speed at which you are travelling.

The second equation is also on page 10 without explanation. It is the speed at which light moves perpendicular to the rod travelling along the X axis, and while he mentions the Y and Z axes he had not explained that the light ray was not a tight beam, but a spherical transmission, meaning that observers on the Y and Z axes would also intercept the light by travelling along one side of the triangle. So he was resolving a triangle, such that the hypotenuse is the Speed of Light, c - the x side has the speed, v, you are travelling at with the rod, and so the third side becomes the usual square of the hypotenuse equal to the sum of the squares of the other two sides, or a vector triangle, so the resulting speed along the y or z axis is the square root of $(c^2 - v^2)$. I did not cotton onto this hypotenuse argument at all when I first read this equation because I wondered for some time what other equation it had come from, there being no diagram to help argue his point. He had just spirited it out of the air!

Thus Einstein was not good at helping persuade others of his argument. After consulting a mathematician to help, his equations finish up on page 16 with

Observer's Time Ћ = Rest Time T multiplied by the square root of $(1 - v^2/c^2)$

It can be seen that Observer's Time actually stops at zero when the Observer is travelling at the speed of light, $v = c$: but we will never do so of course. This equation is the reciprocal of the Lorentz Factor derived by Hendrik Lorentz in 1892.

Physicists in the end decided that Einstein was quite right in his perceptions of the physics of high speed moving bodies, and they have proved the equations many times since 1905. This is the benefit of theoretical physics which Einstein was proposing with his mathematics colleague – once the prediction is known it can be tested.

With respect to travelling to the stars, one million miles per hour is only 0.15% of the speed of light, so when you apply Einstein's famous square root equation for time shown above, you get a time reduction for the traveller relative to us here on Earth of only 0.0001% - obviously insignificant, so you will age just as quickly!

Paper 4 "Does the Inertia of a Body Depend on its Energy Content" also published in 1905, derives $E=mc^2$ in 4 pages.

STEPHEN W. HAWKING

I have read Professor Stephen Hawking's book ISBN 0-593-01518-5 "A Brief History of Time" and found the first half enjoyable and very well and simply written, but I did not enjoy the second half which I found was too fantastical for me, since it was speculating about the evolution, birth and development of the Universe in very esoteric terms, and I have not been immersed in Particle Physics in such depth in my life – at the time of my 1960 A level we did not know of such things. He produced very many papers, including his Reith Lecture. However, I strongly dispute that there is such a thing as negative energy which he talks about and which I now think is what is called Hawking Radiation – I can understand positrons and electrons having opposite charges, but they both have a positive mass and exhibit positive energy.

I was surprised he did not talk about the nature of the black hole singularity itself. It seems to me that the mass of what I call a quark star (the black hole singularity) will not allow light to escape from it at all, because the gravitational effect would be so enormous that light could not escape from it, so the event horizon of the black hole itself would be the mass surface (horizon). However, outside the singularity mass, would be a region where gravity was still so huge that light or radiation from another nearby star, grazing past the black hole, would not be able to escape – so the event horizon in that circumstance would be at an altitude above the quark star surface, I think. It is beyond this region, above the mass of the quark star and above the altitude where gravity has dropped to the extent that light and energy can actually escape, where the overall event horizon exists. I may not have interpreted him correctly, but that is what I think, for Stephen did not explain it at all!

I still have a copy of my A level Physics exam papers, including the Scholarship level paper which I did not take, and none of them include anything about Particle Physics – the 1961 A level Physics (Nelkon and Parker) book I bought (it is a reprint of the 1958 book I actually used, but I cannot find a copy of that) had one short section titled "the beginnings of atomic physics". It seems to me that A level Physics these days concentrates on nuclear power and particle matters, almost to the exclusion of the very wide range of Physics material that I had to learn as shown in the book and the exam papers – mechanics, properties of matter, heat, light, sound, electricity and magnetism.

CHAPTER 23

CLIMATE CHANGE

I want to see a green and fair recovery from COVID-19 that limits warming to 1.5 degrees C or below - this is an hourly average throughout the year, night and day, winter and summer and means even more extremes of temperature. Warming is already happening and the strong evidence is in the Polar Regions where ice is melting at enormous rates and temperatures have reached their highest levels in Siberia. The melting of ice means there is less reflection of the solar heat because the oceanic waters will absorb the heat and make the situation worse. Our climate has become warmer in the UK too and freak weather is becoming more abundant. It may even turn the Gulf Stream away from us.

The only vector that we can manipulate to reduce global warming (no matter its cause) is to reduce our CO_2 emissions – we have no other option, for there is nothing else that we can do. To say there is no such thing as global warming is to ignore the facts, not only of the scientific argument, but particularly of the Arctic and Antarctic regions.

Our best chance at rebuilding a resilient economy goes hand in hand with tackling climate change and restoring nature. The best way to stimulate the economy is by investing in green sectors like clean transport networks and renewable energy. This investment can build our resilience to climate change, create economic opportunities in industries of the future, and offer economic returns which are three to eight times greater than the original investment. It can also deliver wider benefits to the UK of £90bn a year, for example by cutting air pollution and saving healthcare costs. It is crucial that we rebuild, level up and invest in a way that delivers these benefits - for people and for our planet. Therefore it must: power up green investment, deliver zero carbon homes fit for the future, unleash the potential of renewable energy, invest in a clean transport future, boost the recovery of nature in the UK and ensure our international finance supports net-zero and climate justice. Before I outline the key areas that I want prioritised to forge a green recovery, I first want to state my support for the adoption of a 'net-zero test.' This would mean that all spending and Treasury decisions ensure the UK gets on track to net zero as soon as possible. This would help to build a recovery which is resilient and sustainable. Such a rule would be a world first for an economic ministry, showing global leadership ahead of the COP 26 climate summit, which the UK will be hosting in 2021.

During lockdown I have found my local green space to be crucial to my health and well-being. We must celebrate the beauty of our countryside and the diversity of our wildlife by preserving and creating green spaces so everyone can continue to enjoy an essential part of what makes the UK so unique.

Therefore, I support The Climate Coalition's recommendations that the UK Treasury use the coming UK Comprehensive Spending Review to introduce a package of investment that brings together the need to get on track to net zero and boost our natural assets. This would allow the UK to flourish through new employment, new green spaces, cleaner air and waters and secured food supplies. Investment priorities in the UK will deliver these benefits:-

- £500 million annual investment in a National Nature Service to employ and train unemployed people in environmental skills, addressing the economic crisis and building a workforce fit for a future green economy. A National Nature Service will bring employment and opportunity to people who need it now, and provide the workforce needed to create the Nature Recovery Network and to green our cities. In the longer term, the training provided by the National Nature Service will fill the growing green skills gap, equipping people for the green jobs of the future.

- £621 million annual investment to enable delivery of the Nature Recovery Network and nature-based solutions for climate change. By creating new habitats, and enhancing and linking up existing ones, the Government can deliver the Nature Recovery Network - a joined-up system of places where plants and animals can live, feed and breed. Funds should be spent on wildlife and carbon rich habitats such as wetlands, peatlands, meadows, saltmarshes, woodlands, sea grass and coastal realignment. These new habitats will be integral to meeting our net zero target and deliver co-benefits like flood prevention and better air and water quality.

- £5.5 billion investment in levelling up access to nature in our towns and cities so that every community can access green infrastructure, high quality green spaces, and the health and well-being benefits they provide. By upgrading existing parks, creating new ones, and planting green streets and urban forests we can offer millions of people access to explore and enjoy wild, natural spaces, without the need of a car. The impact of this access will deliver £200 billion in health benefits to urban communities over 30 years, a return on investment of 40:1

- £2.9 billion annual investment to develop world-leading food, farming and fishing sectors, with high standards. The financial assistance schemes the Agriculture Bill will introduce, based around the tenet of public money for public goods, must protect our soils, seas and pollinators, enhance animal welfare and secure a long term future for the farming and fishing sectors. Although, the schemes will not be available to all before the end of the next Spending Review period, early benefits could be accrued by a transitional programme, which would help to prepare farmers and others for the new schemes and create a ladder to enable these sectors to increase their ambition and deliver good environmental outcomes swiftly.

- £118 million annual investment to boost jobs in marine and coastal protection, including £70 million per year for the expansion, protection and management of the UK's entire network of Marine Protected Areas, including monitoring and enforcement costs, to ensure that MPAs covering at least 30% of UK waters are fully or highly protected by 2030; £9 million over the next three years for the establishment, monitoring and maintenance of seagrass, saltmarsh, mudflat and oyster bed restoration sites; and confirming £45 million per year of funding for sustainable fisheries management from the end of the post-Brexit transition period.

- £200 million investment in world-leading regulation, data-science and information systems and expertise. To enhance our ecosystems, we first need to understand them. This can only be achieved with comprehensive data that provides quality information about species and habitats and evidence of the value of different ecosystem services. Investment in DEFRA's delivery bodies—especially Natural England, the Environment Agency and the Office for Environmental Protection—is also needed to ensure that advice, monitoring, enforcement and regulation can be carried out effectively. This package of investment will bring nature back to health, and us with it.

I want to be able to heat and power my home with energy that doesn't pollute the air or contribute to climate change. Costs of renewable power - especially wind - have already fallen, and it's a win-win to invest in clean energy generation that is better for our planet and leaves more money in our pockets.

Therefore, I support The Climate Coalition's recommendations:

- Build on the Government's manifesto pledge to deliver 40GW offshore wind by 2030, capitalising on the success of the renewables sector to ensure that overall generation is tripled by 2030. Now that costs have fallen so far, this will be an essential way to lower energy bills. I note that nothing is said about creating more energy storage, so essential to reduce our reliance on gas turbine electricity generators - huge lithium-ion batteries are so essential.

- Auctions for onshore wind, solar and offshore wind should be held every year to ramp up deployment. Investment should also be increased in skills training, supply chain expansion and port infrastructure.

- Planning must also be improved to optimise opportunity for low cost power whilst avoiding harm to wildlife or special landscapes. Rooftop solar should be supported through increased procurement on public sector roofs and fiscal incentives, such as lower business rates to encourage private investment.

- Set up and capitalise a Climate Infrastructure Bank - which can play a leading role in harnessing the opportunities arising from investing in the industries and technologies of the future, increasing prosperity across the UK's regions and leveraging the additional private finance that will be needed. As the UK loses access to the European Investment Bank, now is the time to race to the top with a world leading Climate Infrastructure Bank that drives innovation, breaks down the barriers to low carbon investments and supports long-term economic productivity. It should include a green banking function to de-risk and aggregate zero carbon infrastructure projects; support for the development of future markets in zero carbon technology by investing in innovation; provision of strategic advice to the government on the barriers to green investment and how to address them; an international investment arm to leverage in major funding to clean sectors around the world and support nature-based solutions. A team should be appointed to lead the detailed design and implementation work over the course of 2021.

- Adopt and implement a Net-Zero Finance Plan - a net zero strategy can only be effectively implemented if the Government also has a plan to finance it and maximise the leverage of private investment across every sector of the economy. The plan must involve comprehensively assessing the potential role of regulation, as well as tax and spending options to deliver the transformation required in each sector over the long term. To include how to finance decarbonisation of transport, power, industry, agriculture, nature based solutions and buildings.

- Putting private finance sector on track to net-zero and 1.5C, we are asking the Treasury to make climate risk disclosure mandatory, and to require all UK regulated financial institutions to have a strategy in place by the end of 2021 to meet the Paris Agreement, extending across global practices. The UK financial sector is one of the centres of global fossil fuel finance and must transition to a net-zero future along with the rest of the economy if it is to remain globally competitive.

Failure to invest in a timely and well-managed net-zero transition will come at a huge cost to the UK economy. However, an investment gap exists between what is planned, and what is required to decarbonise the UK's infrastructure.

The UK has shown leadership in supporting disclosure of climate change risks and opportunities by private sector firms, but it is now time to move from disclosure to action. Moving a step forward by requiring financial firms to publish their transition plans for achieving net zero financed emissions will leverage change throughout the economy as investors, lenders and insurers engage with firms in the real economy and set expectations of decarbonisation.

The UK has a vital role to play in creating a greener, healthier and fairer future overseas, as well as at home. As a global centre of finance, and with international financial budgets, it's important that these contribute to a future-proofed tomorrow and climate justice:

- Put an end to public finance for fossil fuels overseas, as the logical next step from its previous pledge to end finance for coal overseas and its leadership in building an international coalition to 'power past coal'. This would mean ending support - including ODA and UK Export Finance - for all fossil fuels (coal and oil) overseas, reviewing and phasing-out existing investments by the end of 2021, and scaling up support for renewable and efficient energy systems and for universal access to energy. The UK should focus all its support on assisting countries in the Global South to transition or leapfrog to renewable and efficient energy, and delivering on universal energy access.

- The UK Government needs to maintain its commitment to provide £11.6 billion over five years (2021/22 to 2025/26) for climate finance, and within that maintain the UK's commitment to 50% of this finance being for adaptation. International climate finance is a cornerstone of global cooperation on climate change, and will be central to a successful COP26 outcome in 2021.

- Assign additional funding to support vulnerable countries to pursue low-carbon development pathways, adapt to climate change, and manage loss and damage. No new and additional (to pre-existing aid commitments) sources of finance have yet been provided for developing countries to tackle climate change, despite historic promises of this from the UK. The economic impacts of the COVID-19 pandemic make this challenge all the more urgent and also open a window of opportunity for action that will create long-term impact by supporting clean growth pathways.

- Government spending should also be increased, in line with the UK's fair share, to support global efforts to protect and restore nature. These funds must also be new and additional sources of public finance and must not come from the 0.7% commitment to aid.

- Clearly also we need to insulate our homes up to at least EPC C category to reduce energy bills. That means loft insulation and cavity wall insulation. Those houses without cavity walls should have wall drylining.

PLASTIC

The environment is something we humans are destroying – the seas are full of plastic which is killing our whales and dolphins, let alone ourselves when we eat the fish with all those micro-plastic beads inside them. Micro-beads come from cosmetics we all know, but cigarette filter butts leave micro-plastic remains on our beaches in huge quantity too, all of which gets into the sea and contaminates our food chain. We should stop people dropping cigarette butts on our beaches. If the young whales and dolphins are killed we will lose them all in a generation, for you need offspring to propagate a species. This is a real danger, and we need to stop disposal of all of that plastic into our rivers and oceans. We need some way of collecting all of those jellyfish-like plastic bags that gorge their stomach and intestines – with nets? But how? The oceans are so deep! Maybe we can clear the shallower waters….. And the micro-plastic has to be stopped in 2023, for there is no way of netting them, and what is already waste in our oceans and rivers will never be reduced for it will recycle through our food for several generations – it is a crime in my opinion that this stuff was ever allowed to be manufactured/sold! Why did Trading Standards or the Health and Safety Executive not pick this up before it was allowed onto the sales market? Surely it ought to be part of their job, from now on at least, for all plastic manufacturers new sales.

The Prime Minister has just announced a 25 year prevent plastic campaign - I fail to understand why this is not 5 years, for we will find it difficult to argue with Indonesia, India and China (for example) that they should rapidly stop their dumping of waste into rivers and oceans with a 25 year argument! We need to care far more for our planet, not dump yet another 25 years worth of plastic, for this is a crime which is potentially worse than climate change.

I have heard that the waste we send to places like China, is NOT properly treated. They recycle the easy plastic, but then dump the rest into their rivers – thus it is our waste that is in Chinese rivers, and thence our Oceans, as well as their own. We must stop exporting it to China and deal with it ourselves, and I am pleased that China have now refused it. A company called Recycling Technology Ltd in Swindon, has come up with a new home grown solution, which effectively turns plastic back into the sort of oil from which it was produced in the first place, and they claim that overall they can recycle 90% of our plastic waste – including crisp packets! It will take time to manufacture all of the machines that will be needed, and they are only onto their second machine at the moment, but the promise is there.

ACKNOWLEDGEMENTS

Acknowledgements are due to Rolls Royce and MoDUK for the memories that are here portrayed, and to the staff at Westwood Books who so helped produce this book.

REFERENCES

ISBN 978-0-19-923085-3 "Why Evolution is True"

ISBN 0-340-50634-2 "Neurophysiology"

ISBN 978-0713721386 "Ships in Bottles: A Modeller's Guide"

ISBN 978-0-7509-6841-6 "Testing Tornado"

ISBN 978-0-7524-5936-3 "Tornado ADV"

ISBN 0-9530345-0-X "The Birth of Tornado"

ISBN 0-470-01821-6 "Crystals and Crystal Structures"

ISBN 3-540-19747-8 "Parallel Processing for Jet Engine Control"

ISBN 1-86058-289-3 "Aircraft Systems"

ISBN 978-0-9573755-4-3 "EAP The Experimental Aircraft Programme"

ISBN 978-0-7509-6843-6 "Testing Tornado" by Dave Eagles AFC

ISBN 0-201-08329-9 "Lisp"

ISBN 0-00-771883-7 "Longitude"

ISBN 085950-062-4 "Positional Astronomy and Astro-Navigation made easy"

ISBN 978-1979789080 "50 Shapes of Chinese"

ISBN 978-90-411-3730-2 "Achieving the Single European Sky"

ISBN 978-1-84885-506-9 "Power Politics"

ISBN 1-85410-823-9 "Broken Rails"

ISBN 0-7484-0892-4 "Basic Superfluids"

ISBN 978-9810232856 "Microstructures and Structural Defects in High- Temperature Superconductors"

ISBN 0-85498-521-2 "Electric Field Phenomena in Biological Systems"

ISBN 978-0-552-77807-7 "Life on the Edge, The Coming of Age of Quantum Biology".

ISBN 978-0-309-26175-3 "In the Light of Evolution, Volume VI, Brain and Behaviour"

ISBN 978-0-19-852851-7, "How Homo became sapiens – on the evolution of thinking"

ISBN 978-0-349-42546-7 "The Little Book of Humanism"

ISBN 0-521-47168-0 "Reaching into Thought – the Minds of the Great Apes"

ISBN 0-691-12228-8 "Einstein's Miraculous Year"

ISBN 0-593-01518-5 "A Brief History of Time"

ISBN 978-1-78607-718-9 "The Good State – On the Principles of Democracy"

ISBN 0-7506-3917-2 "Engineering Materials Technology"

ISBN 0-340-71920-6 "Strength of Materials and Structures"

"Fluid Mechanics for Engineers" by P S Barna, Butterworths, 1957

BBC4 television "Prehistoric Autopsy"

"A note on the effect of auditory stimulation on the behaviour and welfare of zoo- housed gorillas" by Deborah Wells, Dwyer Coleman, and Mark Challis 2005

The CAA's "Future Airspace Strategy for the UK from 2011 to 2030"

The CAA's "State Safety Programme for the United Kingdom"

The US Office of the Secretary of Defense "Nuclear Posture Review" 2018

The UK SDSR "Strategic Defence and Security Review" 2015 – the next SDSR is due in 2020

Made in the USA
Monee, IL
04 November 2026

67380058R00154